PEOPLE and PIXELS

Linking Remote Sensing and Social Science

Diana Liverman, Emilio F. Moran, Ronald R. Rindfuss, and
Paul C. Stern, *Editors*

Committee on the Human Dimensions of Global Change
Commission on Behavioral and Social Sciences and Education
National Research Council

NATIONAL ACADEMY PRESS
Washington, D.C.   1998

NATIONAL ACADEMY PRESS • 2101 Constitution Ave., N.W. • Washington, DC 20418

NOTICE: The project that is the subject of this report was approved by the Governing Board of the National Research Council, whose members are drawn from the councils of the National Academy of Sciences, the National Academy of Engineering, and the Institute of Medicine. The members of the committee responsible for the report were chosen for their special competencies and with regard for appropriate balance.

This report has been reviewed by a group other than the authors according to procedures approved by a Report Review Committee consisting of members of the National Academy of Sciences, the National Academy of Engineering, and the Institute of Medicine.

The project that is the subject of this report was funded by the U.S. National Aeronautics and Space Administration through Grant Number NAGW-4939. Additional support came through Contract No. 50-DKNA-7-90052 between the National Academy of Sciences and the National Oceanic and Atmospheric Administration, which supports the activities of the Committee on the Human Dimensions of Global Change with funds contributed by the consortium of federal agencies that support the U.S. Global Change Research Program. Any opinions, findings, conclusions, or recommendations expressed in this publication are those of the authors and do not necessarily reflect the view of the organizations or agencies that provided support for this project.

People and Pixels: Linking Remote Sensing and Social Science is available for sale from the National Academy Press, 2101 Constitution Avenue, N.W., Box 285, Washington, D.C. 20055. Call 800-624-6242 or 202-334-3313 (in the Washington Metropolitan Area).

The National Academy of Sciences is a private, nonprofit, self-perpetuating society of distinguished scholars engaged in scientific and engineering research, dedicated to the furtherance of science and technology and to their use for the general welfare. Upon the authority of the charter granted to it by the Congress in 1863, the Academy has a mandate that requires it to advise the federal government on scientific and technical matters. Dr. Bruce M. Alberts is president of the National Academy of Sciences.

The National Academy of Engineering was established in 1964, under the charter of the National Academy of Sciences, as a parallel organization of outstanding engineers. It is autonomous in its administration and in the selection of its members, sharing with the National Academy of Sciences the responsibility for advising the federal government. The National Academy of Engineering also sponsors engineering programs aimed at meeting national needs, encourages education and research, and recognizes the superior achievements of engineers. Dr. William A. Wulf is president of the National Academy of Engineering.

The Institute of Medicine was established in 1970 by the National Academy of Sciences to secure the services of eminent members of appropriate professions in the examination of policy matters pertaining to the health of the public. The Institute acts under the responsibility given to the National Academy of Sciences by its congressional charter to be an adviser to the federal government and, upon its own initiative, to identify issues of medical care, research, and education. Dr. Kenneth I. Shine is president of the Institute of Medicine.

The National Research Council was organized by the National Academy of Sciences in 1916 to associate the broad community of science and technology with the Academy's purposes of furthering knowledge and advising the federal government. Functioning in accordance with general policies determined by the Academy, the Council has become the principal operating agency of both the National Academy of Sciences and the National Academy of Engineering in providing services to the government, the public, and the scientific and engineering communities. The Council is administered jointly by both Academies and the Institute of Medicine. Dr. Bruce M. Alberts and Dr. William A. Wulf are chairman and vice chairman, respectively, of the National Research Council.

Contents

Preface

1   Linking Remote Sensing and Social Science:
The Need and the Challenges   1
Ronald R. Rindfuss and Paul C. Stern

2   A Brief History of Remote Sensing Applications, with
Emphasis on Landsat   28
Stanley A. Morain

3   "Socializing the Pixel" and "Pixelizing the Social" in
Land-Use and Land-Cover Change   51
Jacqueline Geoghegan, Lowell Pritchard, Jr.,
Yelena Ogneva-Himmelberger, Rinku Roy Chowdhury,
Steven Sanderson, and B. L. Turner II

4   Linking Satellite, Census, and Survey Data to Study
Deforestation in the Brazilian Amazon   70
Charles H. Wood and David Skole

5   Land-Use Change After Deforestation in Amazonia   94
Emilio F. Moran and Eduardo Brondizio

v

6 Land-Use/Land-Cover and Population Dynamics,
   Nang Rong, Thailand   121
   Barbara Enwistle, Stephen J. Walsh, Ronald R. Rindfuss,
   and Aphichat Chamratrithirong

7 Validating Prehistoric and Current Social Phenomena
   upon the Landscape of the Peten, Guatemala   145
   Thomas L. Sever

8 Extraction and Modeling of Urban Attributes Using
   Remote Sensing Technology   164
   David J. Cowen and John R. Jensen

9 Social Science and Remote Sensing in Famine Early Warning   189
   Charles F. Hutchinson

10 Health Applications of Remote Sensing and Climate Modeling   197
   Paul R. Epstein

APPENDICES

A An Annotated Guide to Earth Remote Sensing Data and
   Information Resources for Social Science Applications   209
   Robert S. Chen

B Glossary   229
   Mark Patterson

Biographical Sketches of Contributors and Editors   237

Preface

The U.S. government has been collecting data about the earth's surface and atmosphere from planes and satellites for decades. In the past, technology and routines of data collection and management have been developed primarily for the earth science community. More recently, the National Aeronautics and Space Administration (NASA) has been paying increased attention to the potential value of remote data for other users, including social scientists; farmers; local government officials; and land-use, urban, and coastal planners. Remotely sensed data have potential scientific value for the study of human-environment interactions, especially land-cover and land-use change (LUCC), a research area that is now the focus of a joint international research effort of the International Geosphere-Biosphere Programme and the International Human Dimensions Programme on Global Environmental Change. Remotely sensed data also have potential scientific value for addressing questions in other areas of social science, including urban studies and human population dynamics, and practical value for providing predictive information about socially significant events, such as famine and epidemics of disease.

As part of an effort to realize more of the apparent potential of remotely sensed data, NASA asked the National Research Council (NRC) to organize a workshop that would bring social scientists who have tried to use satellite data together with experts in remote sensing technology. The participants were to discuss the lessons of success and failure that have emerged from efforts to link remote sensing and social science, and to identify ways of making satellite observations more useful sources of data for social science, research on human-environment relations, and other applications. The workshop was organized by the NRC's Committee on the Human Dimensions of Global Change and held on November 12-13, 1996.

The workshop included presentations on a variety of applications of remote sensing, as reflected in this volume. More than 50 participants—including social scientists; remote sensing specialists; experts in geographic and data information systems; and staff members from NASA, four other federal agencies, the White House, and the World Bank—engaged in lively discussion about the presentations, the opportunities created by remotely sensed data for social science and related fields, and the limitations and pitfalls that can be encountered in trying to exploit those opportunities.

This volume includes revised versions of most of the presentations made at the workshop, as well as two overview chapters that identify major conceptual, methodological, and organizational issues faced by those who attempt to make greater use of remote sensing for social scientific and related purposes. In addition, it provides a guide to information resources and a glossary designed to facilitate interaction and understanding between social scientists new to the field of remote sensing and remote sensing specialists unfamiliar with social science.

The committee believes this book will make remotely sensed data more accessible and its potential uses more evident to researchers in the human dimensions of global change and eventually to the broader social science community. We hope it will also encourage and sharpen the dialogue between these potential users of remotely sensed data and those who collect and transform the data for scientific use, thus making the data even more useful over time.

On behalf of the NRC's Committee on the Human Dimensions of Global Change, I thank Nancy Maynard of NASA, who first asked us to develop this project. I also thank my colleagues who collaborated with me in organizing the workshop and this volume—Emilio Moran, Ronald Rindfuss, and Paul Stern. The four of us were equal partners in this effort, a fact incompletely reflected by the listing of our names alphabetically on the title page. I wish to thank as well Eugenia Grohman and Rona Briere, who provided essential help in editing and producing the volume, and especially Heather Schofield, whose efforts were essential to organizing a successful workshop, collecting the written contributions, and getting the volume ready for publication.

Diana Liverman, *Chair*
Committee on the Human Dimensions of Global Change

PEOPLE and PIXELS

1

Linking Remote Sensing and Social Science: The Need and the Challenges

Ronald R. Rindfuss and Paul C. Stern

There is increased interest today in making scientific progress through the use of remotely sensed data[1] in social science research. Space-based sensors are scanning the earth's surface and sending back images with increasingly high spatial, spectral, and temporal resolution, and data likely to become publicly available within the next year promise to show considerably improved resolution.[2] Government agencies that collect remotely sensed data, such as the National Aeronautics and Space Administration (NASA) and National Oceanic and Atmospheric Administration (NOAA), have a growing interest in making these data useful to social scientists, and the increased availability of funding for research on the human dimensions of global change provides incentives for social scientists to study human activities with a strong spatial component, such as land-use transformations. This confluence of events sets the stage for social scientists to use remotely sensed data and for social scientists and remote sensing experts to collaborate.[3] This volume examines the potential for such use. It offers some guidance for researchers and research sponsors in the form of reports of promising research, information on the state of the technology, and reflections on the challenges of linking social science and remotely sensed data.

Remote sensing is not a new technology. Aerial photographs have been in widespread use for a half-century (Carls, 1947) and satellite images for a quarter-century (e.g., Estes et al., 1980; Morain, in this volume). These images have been put to various socially useful purposes, including making crop forecasts, predicting severe storms, and planning land development. Despite the apparent usefulness of remotely sensed data for social purposes, however, remotely sensed images have not been a popular data source for social science research, for several reasons.

First, the variables of greatest interest to many social scientists are not readily measured from the air. Many social scientists find visible human artifacts such as buildings, crop fields, and roads less interesting than the abstract variables that explain their appearance and transformations. Changing land use, road and building construction, and the like are regarded as manifestations of more important variables, such as government policies, land-tenure rules, distributions of wealth and power, market mechanisms, and social customs, none of which is directly reflected in the bands of the electromagnetic spectrum. Thus social scientists are likely to be skeptical that remote sensing can measure anything considered important in their fields of study (Turner, in press).

A related issue is that social science is generally more concerned with why things happen than where they happen (Turner, in press). Even areas of social science in which one might expect a spatial orientation are curiously aspatial. For example, while it seems almost self-evident that spatial propinquity must be a factor in the shaping of social networks, it is only recently that the spatial aspects of social networks have been receiving attention (Faust et al., 1997). Relatively few social scientists outside the field of geography value the spatial explicitness that remotely sensed data provide, nor do the typical social science data sets contain the geographic coordinates that would facilitate linking social science data and remotely sensed data.

Further, the scientists who participate in developing remote sensing techniques have overlapped little with social scientists in their backgrounds, theories, methods, jargon, or epistemological approaches, although this situation is changing. Integrating social science and remote sensing will require the fusion not only of data, but also of quite different scientific traditions. Many social scientists do not know what a pixel is, and few have ever considered how clouds may affect data quality. Similarly, the average remote sensing expert is unlikely to be conversant with a wide range of social science problems and solutions, such as why fixed-effects statistical models were developed. It is easy for scientists on one side to underestimate the difficulty of learning the approaches, theories, methods, and jargon of the other. This difficulty is compounded by the fact that those on each side are likely to have some familiarity with the other. Social scientists are likely to have been watching news and weather reports for decades, acquiring what they think is an ability to interpret satellite imagery. Some of the images that appear on the television screen bear a close resemblance to familiar objects, such as maps, making interpretation seem easy. But in fact, these processed images can be several steps removed from the remotely sensed data on which they are based. People who see only these products may have little appreciation of the analysis necessary to produce them. Similarly, socioeconomic patterns and trends are discussed frequently in the mass media. It is easy for those not trained in social science to claim some understanding of it and to think that incorporating it into their research would be straightforward. But as with remote sensing, popularized presentations can mask the detailed analysis that lies behind the summary data. We return to the issue of different scientific traditions later, in discussing the problem of training future scholars.

Finally, bridging the social science and remote sensing fields undoubtedly entails the risks frequently encountered by those who do interdisciplinary research. For example, when we discuss the issues raised in this chapter with remote sensing experts who were originally trained in the social sciences, they frequently mention feeling marginalized from their original fields of study because the problems and concepts central to their remote sensing research are not considered core to those disciplines. Estes et al. (1980) have discussed this issue from the perspective of geography—the discipline in which social science and remote sensing have most closely converged—and we are told that the situation remains much the same in the late 1990s.

Given this gap between the social sciences and remote sensing, why bother trying to bridge the two fields? The question has different answers for different kinds of scientists. For some remote sensing experts, a compelling answer is social utility: remote sensing is expensive, and government spending on it is more justifiable if it improves our understanding of the social system by being incorporated into social science research. While remotely sensed data have been employed for a variety of socially useful purposes, such as increasing yields through precision farming or weather forecasting, there are, as noted, relatively few examples of those data being used in social science research. The social utility argument posits that remote sensing becomes even more valuable to the extent that social scientists find it useful, and that efforts should be made to identify and overcome the barriers to making this happen. In addition, the contributions of social scientists might allow remote sensing experts to "see" landscape features in the remotely sensed data not previously apparent. There are several examples of this in the present volume.

From the perspective of social science, one important reason for using remotely sensed data is to gather information on the context that shapes social phenomena. The role of context has been central to the theories and empirical work of numerous sociologists, economists, and anthropologists. Remote sensing offers an additional source of contextual data for multilevel analyses. Another consideration involves the growing interdisciplinary community of scientists interested in sustainable development, pollution prevention, global environmental change, and related issues of human-environment interaction who need to compare data on social and environmental phenomena at the same spatial and temporal scales. This community includes both social and physical scientists. For them, fusing social and remotely sensed data should be an attractive strategy.

Although linking of remote sensing and social science is difficult, it has been and continues to be done, as evidenced by the case studies presented in this volume. This chapter examines why it is important to join people and pixels, addressing some of the challenges of doing so. Understanding the challenges is essential if progress is to be made. In considering the promise and opportunities offered by collaboration between remote sensing and social science, we want to strike a balance between overpromising and underselling. We do not believe remote sensing will quickly revolutionize social science; rather, we suggest that some progress can be made by joining social science and remote sensing perspectives, techniques, and data. Hence, the majority of this volume consists of examples of the use of remotely sensed data—mainly from space-based platforms—in social science research.[4] However, we do not want to be overly constrained by the present, so we also speculate about additional, as yet untried, applications of remote sensing to social science questions.

The volume is intended to stimulate dialogue between social science and remote sensing experts, and in any dialogue, it helps to know something about the participants. The Committee on the Human Dimensions of Global Change at the National Research Council, which is responsible for this volume, consists of social and natural scientists interested in the scientific understanding of human-environment interactions. The majority of the committee members were trained in the social sciences, and the authors of this chapter are typical in this regard. One of us (Rindfuss) is trained in sociology, with research and teaching interests in population studies or demography. He began graduate school wanting to understand American fertility patterns and trends, and then branched out geographically and substantively, while continuing to publish on fertility in the United States. He has used remotely sensed data in work on population migration and social change in Nang Rong district, Thailand. He has worked mainly with micro-level data sets in which individuals or households are the units of analysis. He has been involved in various interdisciplinary activities through the Population Association of America, the National Research Council, and other multidisciplinary organizations and has directed an interdisciplinary research center, but approaches the topic of people and pixels from the vantage point of a sociologist/demographer.

The other author (Stern) is trained in social psychology, but has long been interested in human-environment interactions, particularly in behaviors at the individual and household levels that affect the use of natural resources and the generation of waste and pollution. He has worked with data on individuals' attitudes, beliefs, and behavior and has studied the effects of interventions aimed at changing environmentally significant behavior at the local and regional levels. He has published research together with colleagues from various disciplines in the social and natural sciences, but despite his experience in interdisciplinary collaboration and the potential of remote sensing to provide data on the environmental effects of social interventions, he has not used remotely sensed data in his research.

WHAT CAN REMOTE SENSING DO FOR SOCIAL SCIENCE?

One rationale for linking people and pixels is that doing so might result in better social science research. This could happen in several ways, although the realistic potential for making these improvements is in some dispute.

Measuring the Context of Social Phenomena

Many social science theories relate individual or household behavior to the context within which the individual or household is located. "Context" can denote a variety of entities, including a political or administrative unit, a social network, a school, or a racial or ethnic group. When the individual is the unit of analysis, the individual's household is also a context. People live their lives in contexts, and the nature of those contexts structures the way they live. Contexts can provide advantages (for example, growing up within a wealthy school district) or produce constraints (young adults in rural areas with poor soil quality are more likely to out-migrate). Hypotheses from theories of context may involve additive effects (teenagers residing in high-crime neighborhoods are more likely to become involved in crime than are teenagers in low-crime neighborhoods) or interactive effects (the negative effect of education on fertility is stronger for blacks than for whites)—but in either event, the hypotheses concern the effects of context on individuals or households.

Contexts can be measured in various ways. Censuses, because they obtain information on almost all individuals and households in a country, can be aggregated to various units (block, neighborhood, school district, city, county, or state) to provide measures of the demographic or socioeconomic characteristics of those units. The choice of scale and of characteristics depends on the theory and the hypothesis being tested; the effects of the contextual variables are estimated using statistical models. Sometimes individual respondents know the contextual variables well enough to provide them directly. Race, ethnicity, and religion are examples. Sometimes contextual variables must be measured with data that are not gathered from individuals. Examples include public expenditures on education and laws governing land tenure. In the case of social networks, researchers are experimenting with several approaches to measuring the structure of networks and positioning individual respondents within that structure.

Remote sensing provides an additional means of gathering contextual data, particularly in describing the biophysical context within which people live, work, and play. First of all, remotely sensed data provide an alternative representation of geographical context to that given by maps. Maps always include the map-maker's selection of what is important to represent, and remotely sensed data, though also imperfect representations of reality, have different biases. They can therefore offer a check on what is in maps, additional information, and sometimes a useful alternative perspective. A good general source of information on methods of measuring and understanding geographical contexts is the recent volume *Rediscovering Geography* (National Research Council, 1997).

In addition, remote sensing has the potential to supplement georeferenced social data by characterizing numerous aspects of the context, ranging from land cover to soil moisture to weather. An example is the work of one of the authors (Rindfuss) in Nang Rong district, Thailand, mentioned earlier. This work (the subject of Chapter 6 in this volume) concerns the determinants of out-migration. The Nang Rong district is a rural, agricultural area, with rice being the predominant crop. Over 90 percent of the adults are farmers, so access to land is essential for young adults seeking employment in the district. Since forested land in Nang Rong tended to have ambiguous legal titles, it was expected that the availability of forested land would reduce the likelihood of out-migration of young adults, and this was indeed found to be the case (Rindfuss et al., 1996). Another example comes from the work of Geoghegan and colleagues (see Chapter 3), who are analyzing the effect of the mosaic of land uses surrounding a property on its economic value and the probability of its future development.

Migration raises one of the thorniest issues related to the use of contextual data to study social processes: individuals and households can change their contexts through migration. People move for many reasons, including better economic opportunities, better schools, and a preferred biophysical environment. When they move and change the context in which they live, that context needs to be modeled as an endogenous variable, rather than a simple external influence on behavior. Even without migration, individuals can act to change their contexts—a possibility that may be more easily uncovered when contexts are measured in interviews than when they are measured by remote sensing. Thus, theoretical care is needed when using remotely sensed data to supply contextual data for models of individual or household behavior.

Measuring Social Phenomena and Their Effects

Remote sensing can provide measures for a number of dependent variables associated with human activity—particularly regarding the environmental consequences of various social, economic, and demographic processes. For example, remote observations of land cover may show the footprints of agricultural extensification, urbanization, and road development;[5] observations of vegetation density may be related to the effects of fertilization, irrigation, and other agricultural practices; and observations of new building construction may be linked to the effects of local policies on land use and property taxation. Remote sensing has sometimes proven to be the best method for identifying archaeological sites and relating them to key features of their geographical settings (see Chapter 7).

Models that combine remote observation with ground-based social data have the potential to improve understanding of the determinants of various land-use changes. Geoghegan and colleagues in Chapter 3 and Cowen and Jensen in

Chapter 8 give examples of such modeling in which residential development is the variable being predicted. It may also be possible to study the effects of changes in agricultural commodity prices on cropping patterns and tillage practices by combining price data with remotely sensed data, and to improve understanding further by incorporating additional data on land-tenure systems or agricultural policies.

Providing Additional Measures for Social Science

Social scientists frequently use aggregated units of analysis: cities or towns, counties or districts, states or provinces, or countries. The substantive questions vary, but investigators typically use multiple indicators for these aggregated units. Remote sensing can provide a variety of additional indicators for these studies, including land cover, moisture measures, locations of major roads and hydrographic features, and indicators of crop fertility. Gathering such measures from the ground might be possible, but often is prohibitively expensive because of the need to collect large amounts of small-scale data for aggregation. Remote sensing can sometimes provide highly aggregated data at less cost.

Indicators from remote sensing can complement indicators from ground-based sources. For example, agricultural intensification can be measured by using data from surveys of farmers' behavior, sales figures on agricultural chemicals and farm equipment, or remotely sensed data on crop density and color. Combinations of social and remote data can yield a deeper understanding of the types of intensification possible, as shown in the analysis of Amazonian agroforestry in Chapter 5. Urbanization can be measured by counting building permits, sampling and observing city blocks, or remotely sensing the proportion of land covered by structures (see Chapter 8). Each data source has its imperfections, but combining sources with different limitations might provide a better picture of the entire phenomenon. In this way, remote sensing—even with its imperfections—can make a contribution to social scientific measurement by improving on some measures and cross-checking others.

Because remotely sensed data are available with greater spatial and temporal resolution than data from other sources, there has been discussion of the potential for using the former data to conduct finer-grained studies than are possible with typical social science data. These possibilities are expanding as higher-resolution data become available from new satellites and satellite data collected by military and intelligence organizations in the United States and the former Soviet Union are declassified. For example, census data are remarkably accurate in most countries, but they are collected infrequently, typically every 10 years. And there are some countries for which census data are not available, and some in which the data are reported inaccurately for a variety of cultural and geopolitical reasons. Some have expressed the hope that remotely sensed data could be used during intercensal periods to update the census reports. Cowen and Jensen (Chapter 8) report on correlations between remotely sensed indicators of dwelling units and actual census counts in data from South Carolina. The validity of their indicators in other geographical and social contexts or over time has not yet been demonstrated, however. There have been some suggestions that remotely sensed data of fine spatial resolution might be used in statistical models to generate estimates of population counts. If that were possible, there would be numerous uses of such estimates. It remains to be seen whether efforts along these lines will yield accurate estimates. Before this becomes possible, however, a number of methodological studies are necessary because of certain inherent limitations in the use of remote sensing for population estimates, such as the inability to sense the number of people per housing unit or housing units per building, or to discriminate between residential buildings and some others. Thus, ground-based studies are necessary to determine how the number of people per dwelling unit varies with the socioeconomic and physical characteristics of neighborhoods. If the variance is sufficiently systematic, remote sensing might help improve intercensal population estimates.

Remote sensing might also help with the census undercount problem. One source of a census undercount is the failure to recognize a physical structure that is a dwelling unit. For relatively remote rural areas, finding dwelling units is a difficult undertaking, and missing a dwelling unit can contribute to the undercount.[6] The use of satellite images with high spatial resolution might improve this process—a possibility the U.S. Bureau of the Census has investigated using aerial photographs (Carls, 1947).

Remotely sensed data have been used for measuring other socially significant variables, especially in urban and suburban contexts. Cowen and Jensen (Chapter 8) describe the use of remote observation to classify land use and land cover into categories established by the U.S. Geological Survey; to measure the area, height, and volume of buildings; to measure traffic patterns and road conditions; to estimate residential energy demand; and to build predictive models of residential expansion. Some of these measurement methods are in the early stages of development, so more experience is necessary to determine how well they work across a variety of social and geographic conditions and over longer periods of time. Nevertheless, these measures may provide important advantages in cost or temporal resolution over conventional measures of the same variables, and may make it possible to improve the quality of modeling used for planning urban infrastructure needs and forecasting the need for utilities or other public services.

It may be argued that remote sensing can support comparative social research—studies that attempt to draw conclusions by systematically comparing the same phenomena in different countries or different regions of the same country—by providing comparable measures of variables these studies need to investigate. Clearly, remote sensing is well suited to providing comparable data for different geographic regions or at different times. The question is whether the parallel social data are available in forms that are comparable.

Making Connections Across Levels of Analysis

Social science disciplines and subdisciplines have their preferred levels of analysis and often do not communicate across those levels. For instance, psychologists and sociocultural anthropologists tend to work with individuals and small groups; political scientists and geographers tend to work at higher levels defined by political units or geophysical features; while sociologists tend to specialize in one level of analysis or another, from individuals to small groups to communities to the world system. Remotely sensed data are essentially global in coverage,[7] composed of individual pixels that can be combined to allow work at any scale or level of analysis more coarse than the pixel size. Thus remotely sensed data offer some potential for encouraging social scientists to think across levels of analysis and to develop theories that link these levels. An example is in the work of Moran and Brondizio documented in Chapter 5 which, starting from an anthropological and highly localized perspective, developed ways of examining land use in geographically disparate areas of Amazonia and thereby addressing regional-level questions. Similarly, Entwisle and colleagues (Chapter 6) collected data on villages in a way that, with the help of comparable data across villages from remote sensing and social surveys, has the potential to place the behavior of individuals in the context of their villages and interpret the characteristics of the village in the context of the region. The issue of linking levels of analysis is explored in more detail in Chapter 3, which suggests some interesting possibilities for combining remotely sensed and ground-based data to study the effects of global economic forces on the behavior of individual land users. Chapter 3 also considers a cousin of the linkage issue in the temporal dimension: the property of path dependence in dynamic systems, which may be affected by their histories as well as their current conditions.

Providing Time-Series Data on Socially Relevant Phenomena

Time-series data can be helpful when social scientists attempt to trace relationships of cause and effect but cannot use experimental methods. Remote platforms sometimes provide time-series data of good comparability (i.e., the same variables measured in the same way across time) on variables of interest to social scientists concerned with the effects of context on behavior or with processes of human-environment interaction. Examples in this volume include data on thinning of forests by human action (Chapter 3), forest regrowth after clear-cutting (Chapter 5), and development of algal blooms that harbor pathogens (Chapter 10). In addition, remotely sensed time-series data can be essential for modeling human-environment interactions. Examples in this volume include the use of remotely sensed data to model the effects of access to forests on out-migration (Chapter 6) and processes of land conversion to urban uses (for example, in Chapter 3).

WHAT CAN SOCIAL SCIENCE DO FOR REMOTE SENSING?

As noted earlier, to the extent that remote observations provide uniquely useful information for social research, these social science applications of remote sensing can be used to provide additional justification for the money spent on observational platforms and data management systems. In addition to this potential practical value of social science to remote sensing, there are several kinds of scientific contributions to remote sensing that might come from its interaction with social science.

Validation and Interpretation of Remote Observations

Remote sensing specialists are well aware of the need for "ground truthing," that is, for validating remote observations against data collected on the ground. A standard example is the problem of measuring land cover by old-growth forests (Lucas et al., 1993; Moran et al., 1994; Skole et al., 1994; Moran and Brondizio, in this volume). It is necessary among other things to distinguish the spectral signatures of old growth from those of forests regrown after deforestation. Doing so requires comparsion of the remotely observed spectral properties of plots known from ground observation to fall in these categories in order to develop an algorithm that accurately discriminates between the two. Further ground observation is required, of course, to validate the algorithm on plots of land not used to develop it.

Although classifying types of land cover requires observations on the ground, it is not usually considered a social science activity. There are, however, some kinds of ground truthing that involve classifying remote observations into more obviously social categories, and thus depend on social science input. An important example is classification of land uses, which are socially defined in ways that do not correspond exactly to categories of land cover. Thus, some tree cover is socially classified as forest land, some as park land, some as suburban landscaping, some as orchard, and some as productive agroforestry land. It is frequently necessary to rely on human informants to make these distinctions.

Similarly, different kinds of land tenure, such as family ownership, village commons, and sharecropping arrangements, may all be used in the same kinds of productive activity and may therefore fall within a single land-cover, or even land-use, classification. It may be possible to associate different management practices that can be distinguished spectrally by remote observation with differences in tenure. Discovering such differences would likely require collaboration between remote sensing specialists who can distinguish spectral patterns and social scientists who can classify land-tenure types and land-management practices.

Data Confidentiality and Public Use

As noted earlier, remotely sensed data are becoming available for public use in ever finer spatial resolution, increasing the ability to discern the footprints of socially important activities. Moreover, as high-resolution military observations are declassified and made available, various organizations will gain access to information about the landscape heretofore not considered by those responsible for the landscape. As improved technical capabilities, collaboration with social scientists, and especially the linking of remotely sensed data with social data make remotely sensed data increasingly useful, new problems and conflicts may arise over the use of the data.

Although there are legal precedents that limit privacy rights with respect to high-resolution aerial photography, the courts have not yet directly addressed questions of privacy and Fourth Amendment rights in the context of space-based remote observation (Uhlir, 1990). As such observation provides improved resolution, new claims of infringements of privacy may surface. There may be new calls for the restriction of access to remotely sensed data, stricter regulations or legislation governing what can be collected remotely, or curtailment of public resources invested in remote sensing technology. There are also unresolved issues of international law (Hosenball, 1990). By way of illustration, when one of the authors (Rindfuss) first showed his Thai collaborators the satellite images for our study site during a research seminar in Thailand, their first question was: "Where did you get these?" Their tone and facial expressions suggested surprise and perhaps a bit of concern that one could simply buy such images. Informal discussions with others using remotely sensed data suggest that the reactions of our Thai collaborators are not unique. Already, some landowners are concerned that remote platforms will reveal secrets about their land-use practices (perhaps revealing to government officials that those practices are violating land-use regulations). We expect that notwithstanding legal precedents in the United States dating from the early days of aerial photography, the increasingly finer resolution and widespread availability of satellite images and their linking to ground-based data sources will fuel public concerns about invasion of privacy and reopen some previously closed issues.

Social scientists have experience in dealing with issues of confidentiality in data collection and dissemination that may be of use to remote sensing specialists. Most social science data collection techniques, from face-to-face interviews to participant observation to the analysis of administrative record forms, require that those providing the information be motivated to remain open and honest, which in turn requires that they trust the researchers to use the information responsibly. There are many examples of censuses, surveys, and other studies in which respondents suspected that the data might be misused and in which, as a result, response rates or data quality was unacceptably low, or studies had to be canceled. Thus, social scientists have come to recognize their collective self-interest in maintaining respondents' confidentiality, and in vigorously dispelling any rumors that confidentiality may not be maintained or that the data will be used inappropriately. As a result, social scientists plan their research carefully so as to assure respondents that their identities will remain confidential, and their information will be used only for statistical analysis. In addition, universities and research institutes use institutional review boards to review research designs and ensure adequate protection for respondents.

Important issues of trust and confidentiality arise with public-use social science data sets, which are increasingly being created in response to the high and escalating cost of collecting large-scale social data. Research sponsors, particularly within the federal government, frequently require that researchers make their data publicly available, and this requirement creates a potential conflict with the need to preserve confidentiality. The conflict has tended to be resolved in two different ways. In the first, which has tended to be used with censuses, individual and household data are aggregated to a sufficient level (counties, for example) so that it is no longer possible to identify the individuals or households that provided the data. In the second approach, which is often used with survey data, all identifiers of the respondent (including geographic ones) are stripped from the data set, and data on the individual respondents are then released.

What are the implications of these practices for remote sensing? If data are aggregated to large geographic units, social data can be linked with remotely sensed data, although the procedures are technically challenging. However, this solution to the confidentiality problem constrains analyses to larger spatial and social scales. There are many research questions for which analysis at the county level and above is appropriate, but there are also many important research questions that require analysis of individual- or household-level data.

If geographic identifiers are removed to allow confidential analysis of individual-level data, the public-use data set cannot be linked to remotely sensed observations or other georeferenced data. Social scientists recognize this problem and have tried various approaches to solving it. Usually, these approaches involve explicit agreements between the data collector and the user that may include legal contracts and the posting of a bond. To date, these agreements have prevented any disastrous breaches of confidentiality, but they are administratively cumbersome and can restrict the research process. To the best of our knowledge, there are as yet no examples linking remotely sensed data to publicly available individual- or household-level social science data sets. The linkages that have occurred, including examples in this volume, involve social science data sets that are not yet in the public domain.

Remotely sensed data that are not linked to social data are less likely to pose problems of confidentiality. However, making remote sensing more useful may raise the incidence and severity of conflicts over confidentiality and access to information. Social science, which has more experience with these conflicts, may be able to offer useful insights and institutional responses to the remote sensing community.

HOW CAN REMOTE SENSING AND SOCIAL SCIENCE IMPROVE UNDERSTANDING OF HUMAN-ENVIRONMENT INTERACTIONS?

An additional argument for better collaboration between remote sensing specialists and social scientists is that such collaboration has been necessitated by a new and important set of intellectual and practical problems: those related to understanding and controlling human impacts on the biophysical environment, as well as anticipating and responding to environmental impacts on humanity. Environmental quality has been a major concern of citizens and policy makers for over a quarter-century, and there is a compelling need to understand human-environment interactions. Such understanding depends on better knowledge of biophysical systems, of human activity, and above all, of the relations between the two. Linking remote sensing and social science is a necessary part of developing this knowledge.

Interpreting, Modeling, and Predicting the Dynamics of Natural Resources

Many projected regional and global environmental problems may be the consequence of human activities that alter land use and land cover (e.g., tropical deforestation and wetland conversion) or affect agricultural productivity (e.g., practices that increase soil erosion). Analyses involving both remotely sensed and social data are critical for understanding these dynamics.

Several contributions to this volume illustrate how remotely sensed and social data can be combined to help us understand human-environment linkages. Wood and Skole (Chapter 4) report their initial effort to predict deforestation, habitat fragmentation, and secondary growth in Amazonia from census data on population dynamics, economic activity, and other social indicators at the regional level. Entwisle and colleagues (Chapter 6) illustrate some of the dynamics of mutual causation between land cover and human migration, with implications for resource demands in both rural and urban areas of Thailand. And Sever (Chapter 7) graphically illustrates the effects of human settlements and public policy on forest cover in the Petén and adjacent areas of Mexico.

Understanding the Human Consequences of Climate Flux

Fluctuating weather patterns concern citizens and policy makers because of the tangible effects they may have on human health and well-being. Research combining social and remotely sensed data can lead to new understandings of the consequences of such fluctuations.

For example, Hutchinson (Chapter 9) describes the development of a famine early-warning system for Africa based on remote observations of drought phenomena, combined with an understanding of the social processes by which people adapt to drought. Generally, forecasts and status reports on food crop growth in drought-prone regions, derived mainly from remotely sensed data, are combined with ground-based data on patterns of human response to generate famine warnings. Such warnings give aid agencies and governments enough advance notice to act to prevent out-migration by threatened human populations.

In Chapter 10, Epstein provides several examples of actual and potential uses of remote sensing to link climatic change and variation to human health through changes in the ecology of disease organisms and their vectors. For example, remote sensing of algal blooms in the Indian Ocean can help provide early warning of cholera outbreaks instigated by disease vectors that feed on the algae (Colwell, 1996). A combination of remotely sensed data and ground observations of rodents helped solve the riddle of a Hantavirus outbreak in the southwestern United States and reveal the linkages among climate variation, ecological change, and human health.

These examples illustrate how remotely sensed data can be used to build forecasting models that have the potential to serve very important social purposes. A significant challenge to social science research is to build understanding of how these forecasting models work and how they can be made more socially useful.

MAJOR INTELLECTUAL ISSUES

We have already noted that combining social and remotely sensed data presents challenges because of dramatic differences in the intellectual traditions that produce and use the two kinds of data. Once these challenges have been met, as they sometime are, by researchers who are willing to make the effort to understand and respect each other's perspectives, several other issues arise to challenge those who would combine the two data sources.

Finding the Appropriate Spatial and Temporal Resolution

Decisions about the appropriate scale, level of aggregation, and frequency of measurement of various data are driven by considerations of both theory and data availability. On the theoretical level, the appropriate units of analysis depend on the question being asked. For example, the debate over global warming rests mainly on questions about what is happening on a global scale and on a temporal scale of decades to centuries. By contrast, questions about population migration turn on the decisions of individuals and households. Some questions require analysis at multiple scales. For example, questions about land use and land cover typically require information at the level of individuals and households that may own the land and make many of the decisions on how it is used, local governments (because they often regulate land use and make decisions about the location of transportation infrastructure), and governments at higher levels. Geoghegan and colleagues (Chapter 3) examine this question of scalar dynamics in more detail.

Data availability also influences the choice of scale or aggregation. With remotely sensed data, the characteristics of the sensing instrument determine the finest grain available spatially and the frequency with which measurements can be repeated. On the social science side, the level of resolution is determined by the nature of the data gathering technique and by any aggregation that may have been done before the data were released in a public-use file. For example, data about the actions of legislative bodies are intrinsically gathered at rather highly aggregated levels, whereas data on educational attainment, even if presented in an aggregated format, necessarily begin with information about individuals.

Researchers deciding to link social and remotely sensed data must make decisions about the appropriate level of aggregation. On the social side, ignoring temporal issues, the finest grain is an individual. Levels of aggregation above the individual depend on the researcher's theoretical perspective, as well as on data availability. Aggregation by political or geographical units progresses through successively higher levels, such as villages, towns, or cities; counties or districts; states or provinces; countries; and regions or continents. Linking such units to remotely sensed data can be relatively straightforward if the units have clear geographical boundaries. However, when aggregation is based on other kinds of theoretical categories, linking to remotely sensed data is more difficult. For example, individuals can be grouped at successively higher levels of aggregation into families, kin groups, and lineages—units that do not always have clear geographic boundaries. Linking such units to remotely sensed or other geocoded data necessarily requires some simplifying assumptions.

On the remote sensing side, the issues of linkage are somewhat different. There is no unit that is comparable to the individual. The smallest unit of observation in satellite data is the pixel, and its size is determined by the measuring instrument, rather than by any theoretically meaningful concept. For example, data from the French Système Pour l'Observation de la Terre (SPOT) have a much finer spatial resolution than Advanced Very High Resolution Radiometer (AVHRR) data. The size of the pixel, in turn, determines the smallest thing one can "see" with satellite data. A large pond may be quite visible with SPOT data, difficult to see with Multispectral Scanner (MSS) data, and impossible to see with AVHRR data. One can aggregate pixels to produce a coarser resolution as dictated by theoretical or substantive concerns, but it is not possible to go below the pixel for finer resolution.[8] (One can, however, combine satellite images with different resolutions or combine satellite images with aerial photographs in an attempt to merge the properties of various remotely sensed images.) In Chapter 8, Cowen and Jensen summarize the pixel sizes and frequencies of observation necessary for measuring a variety of attributes of urban development.

Linking People to Pixels

A critical issue in linking people and pixels is the decision on where to georeference individuals or other social units. In some cases, a social unit has a natural georeferent, but frequently this is not the case. Consider individuals, and assume that the substantive questions being examined involve the effect of individual behavior on some aspect of the land. To which pixel or pixels should the individual be linked? The question can be difficult to answer because, although the land units represented by the pixels do not move, people do.

Researchers usually know their respondents' place of residence because this is typically the location of the data collection effort or because data on place of residence are routinely collected, and the natural tendency may be to link the person to the place of residence. In a setting where the primary economic activity is farming and the primary means of transportation is walking, using an area near the person's residence seems an appropriate way to link people to pixels. This is the assumption used by Entwisle and colleagues in Nang Rong, Thailand (Chapter 6), where people live in clustered villages, rice farming is the predominant occupation, and walking is the main means of transport.

However, as modes of transportation evolve to enable travel over greater distances and as occupations diversify to include manufacturing and service positions, the validity of linking individuals to the pixels where they reside becomes questionable. For example, in the United States in 1990, only 4 percent of the labor force walked to work, and 3 percent worked at home (calculated from U.S. Bureau of the Census, 1993: Table 18). The great majority (73 percent) commuted to work alone in a car, truck, or van. For all those who did not work at home, the mean travel time to work was 22.3 minutes. If their average speed was 48 kilometers (30 miles) per hour, the mean commuting distance was about 18 kilometers (11 miles); if the average speed was 72 kilometers (45 miles) per hour, the mean commute was about 27 kilometers (17 miles). Since people can affect the landscape both at home and at work, one might want to georeference both their residences and their workplaces. Similar considerations apply to shopping patterns, social activities, and religious activities. Routine social science data collection efforts do not provide the capability to georeference these activities; further, tested and accepted methods for collecting such data do not exist.

So far, this discussion has considered only residences and workplaces, and assumed that individuals have just one of each. But many workers have more than one job or work in more than one location. Moreover, increasing numbers of Americans own or occupy vacation homes, and certain geographic areas are being transformed primarily for the purpose of vacationing and tourism by individuals whose primary residences are elsewhere.

An additional problem is that of georeferencing the activities of individuals who participate in the global economy and thereby affect the land in places where they may never have traveled. People buy agricultural and other products from widely dispersed places and manufacture products that produce effluents in the widely dispersed locations where they are used, often without even knowing where their activities are having an impact. Social science is a long way from being able to georeference these human activities, but it is obvious that people in modern economies are not easily linked to pixels for the purposes of understanding the effects of their economic activities.

An analogous problem exists at other levels of analysis. Consider, for example, linking firms to remotely sensed data in order to understand the influence of different types of firms on land-cover or land-use patterns. Should one use only the point locations of a firm's places of business, or should one also consider the commuting patterns of the firm's employees and the locations of its suppliers of raw materials?

One possible approach involves aggregating social data to larger geographical units. This approach assigns individuals to larger areas in which their environmental effects are more likely to be confined. This approach can answer some scientifically important questions and produce interesting results. Examples can be seen in the work of Wood and Skole in Chapter 4. However, there are limitations to this approach. First, there are numerous processes that are not visible at high levels of aggregation. To exploit fully the potential of integrating social science and remote sensing, one should also examine finer-grained relationships. Second, surveys have become much more important than censuses in most social science disciplines, and the majority of surveys are designed to represent a broad geographic area (a country, region, or state) and not to be representative of smaller geographic divisions within that area. This is true even of very large surveys, such as the Current Population Survey in the United States. It is unusual to have comparable surveys conducted in multiple geographic units, and when they are, the units are usually countries, which are too large for many processes of interest.

INSTITUTIONAL ISSUES: MAKING IT HAPPEN

As this chapter and the case examples presented in other chapters of this volume make clear, social science and remote sensing are being linked in efforts to address important scientific and public policy questions. The potential of such collaborations is considerable, although there are also significant challenges. The question for the future is not whether this sort of activity should go on—it will—but how much of it should go on and how it can be facilitated. This section addresses some of the key institutional questions about the future, for example, how to create a productive community of scholars who combine social science and remote sensing, how to train future scholars for participation in this community, and how to support the community with needed data.

Building a Community of Scholars

For the next 5 years, most individuals who work on projects that involve the linking of social science and remotely sensed data will be experts in one of these fields, but probably not both, and will work in collaboration with researchers whose training is complementary. They will acquire some knowledge in the area in which they are not expert, but only the exceptional individual will be expert in both. The volume of research literature in both areas all but precludes an established researcher's becoming an expert in both. Thus in the near term, research combining social science and remote sensing is likely to be multidisciplinary and interdisciplinary, and will involve all the problems associated with such research.

One of these problems is peer review of scientific proposals that incorporate both social science and remote sensing, and that are likely to make contributions to both areas. Ideally, one would want to fund proposals that are of the highest quality and that will make cutting-edge contributions to both the social science and remote sensing components. Social science, however, is a very broad field, and we do not know of anyone who has the audacity to claim expertise in all its subfields. The same is probably true of remote sensing. The peer review process would be best served if it included representatives of all the major subfields of both social science and remote sensing that are represented in the proposals submitted. Given the diversity on both sides, the selection of appropriate reviewers will be a challenging task. Moreover, it may be desirable to fund some projects that break new ground by applying knowledge or techniques that are familiar in either social science or remote sensing in a new and important way, and that are therefore not equally innovative in both fields. Such proposals might be viewed as exciting and innovative by experts in one field, but uninteresting by experts in the other. It will be necessary to find ways of preventing vetoes of such projects by experts in a subfield who do not see the overall value of proposals that go beyond their expertise.

Communication of scientific results is another problem for any new interdisciplinary scientific field. Communication normally occurs through the auspices of scientific associations that are built around disciplines (e.g., economics, geography) or problem areas (e.g., population, natural resource management). The communication media range from small workshops to scientific meetings to journals. To date, there is no scientific association or journal for scholars who are integrating social science and remotely sensed data, and this lack of an institutional base is likely to impede the development of research at the intersection of the two fields. There are certainly examples of sessions at professional meetings that include papers incorporating both social science and remotely sensed data (e.g., the 1996 and 1997 meetings of the Population Association of America, the Pecora 12 conference, and the 1997 Open Meeting of the Human Dimensions of Global Change Research Community). But each of these meetings attracts only a small fraction of those bridging the fields of social science and remote sensing, and further, only one member of a research team usually goes to these meetings (e.g., the Pecora conferences tend to attract only remote sensing experts).

Scientific journals are perhaps the most important communication mechanism. For teams working on projects that use both social and remotely sensed data, the most obvious publication outlets are ones that specialize in only one of the two fields. The peer review process in such journals involves the same problems already noted for the review of proposals: the social science reviewers are generally not competent to review the remote sensing components of the paper, and the remote sensing reviewers are generally not competent to review the social science.

The ability to build a community of scholars depends greatly on publications because they are central to the reward system in science. Universities are the primary employers of social scientists and remote sensing researchers, and they tend to be organized along disciplinary lines. Graduate degrees, faculty appointments, and tenure tend to be determined by disciplinary bodies within a vertical structure in which department chairs report to deans, deans to provosts, and so forth. Uniting social science and remote sensing involves horizontal links across departments or even schools. While these horizontal links are crucial for the research, the fact that they are orthogonal to the decision and reward structure of the university means that tensions will inevitably arise. Even a simple question such as where to publish findings will make collaborators consider their self-interests. Further, many department chairs are wary of interdisciplinary research efforts and centers—they want their faculty spending time and effort in their departments. Although interdisciplinary research is clearly possible, it takes place against resistance in most universities.

The difficulties of peer review, communication, and publication are typical of new interdisciplinary fields, and they are sometimes successfully overcome. Among the ways this has been accomplished are the provision of resources targeted to the field during its developmental period and the orientation of that development around a few applied problems for which there are preexisting communities of scholars. The developing scientific interest in the human dimensions of global change, and within that field the growing attention to and research support for work on land-use and land-cover change, provides a context that can bring together many of the researchers who are combining social science and remote sensing.

Training Future Scholars

As noted earlier, most current researchers who work on projects that combine social science and remote sensing do not have solid skills in both; rather, they are expert in one and collaborate with experts in the other. One model of training would replicate this pattern and add special training of students to be effective interdisciplinary collaborators. A drawback to this approach is that collaboration would increase the expense of research. Another is that collaborations would often be across institutions, creating an issue of distance that would have to be addressed. A third drawback is that graduate students and junior faculty members who were trained and employed in a disciplinary field would have to worry about whether their interdisciplinary research would hurt their chances of getting a job or tenure.

An alternative training model would train young researchers in both a social science and remote sensing. This strategy would address some of the drawbacks of the first model, but has drawbacks of its own. First, it would increase the length of graduate training, which some argue is already too long. Second, such training might have to occur across departments, and most universities are not structured to do such training. Third, young scholars with interdisciplinary training might have difficulty finding employment, especially in universities organized along disciplinary lines.

A third model involves gradual expansion of the community through interdisciplinary research activities that may provide both students and established scholars with what is essentially on-the-job training in remote sensing or social scientific fields that are new to them. This model is an extension of a process that is already occurring in some research institutions. Its chief advantage is that training in the context of ongoing research is likely to be highly effective. However, the process is likely to be slow at first, and it may train people idiosyncratically in narrow segments of a field that are related to a specific research topic.

At this juncture we would not want to recommend one model over another. We anticipate that the models will vary across universities, and depend on the strengths and existing institutional arrangements within each. We would suggest, however, that the time has come for funders, both federal and private, to train the upcoming generation of scholars who could bridge the social science and remote sensing fields.

Providing Necessary Data

Linking remote sensing with social science presents special challenges for data systems. A straightforward but significant problem is to provide georeferencing for social data so as to link them to remotely sensed data, which are normally geocoded. Preexisting social statistics such as those collected by government agencies are typically coded at highly aggregated levels, such as political units. They can be geocoded to some geographic point within the unit, but cannot be disaggregated below the lowest level at which they were made available. [9] If they are geocoded for general use, it is important to select an appropriate geographic point so that researchers can move the point as their scientific purposes require, and to store the geocoded data in an easily accessible place, such as the NASA-supported Socioeconomic Data and Applications Center (SEDAC). New social data could in principle be coded to the location of the individual, firm, farm, or other unit from which they are collected, but as already noted, concerns about privacy and confidentiality often prevent this, and even when it can be done, questions remain about how best to geocode actors in highly interdependent world markets. Again, the geocoded data should be stored, with appropriate documentation, in an easily accessible location. It would be useful to have an organized discussion within the research community about whether it is advisable to develop standard methods of geocoding social data for storage.

Researchers also face the problem of finding appropriate social data to match with remotely sensed data, or vice versa. NASA's support for the SEDAC is intended to address this problem, and an indicator of the SEDAC's success will be the extent to which researchers find it useful for locating the matching data sets they need.

Another challenge is to match the level of resolution of remotely sensed data with that needed for social data. Different social science questions require different levels of resolution in space and time, and perhaps also spectrally. The different needs are illustrated by Cowen and Jensen in Chapter 8 in the context of research on urban dynamics. They are also illustrated implicitly in other chapters. For example, although Chapters 4 and 5 both report studies of deforestation in Amazonia, they rely on data at very different levels of spatial resolution. Each research area probably has its own diverse needs for resolution, depending the scientific questions being asked. This variety of needs in terms of resolution suggests that any standardized system of data storage and geocoding should be highly flexible.

Data maintenance is another important challenge. Many research applications require intact time series of remotely sensed data and benefit increasingly as the time series are extended. For example, improvement of the famine early warning systems described in Chapter 9 depends on continued enhancements to models that account for past data on climate variations, crop production, human response, and famine. The same is true for achieving the promise of public health early warning systems, as described in Chapter 10; for modeling population and land-use dynamics, as described in Chapters 4 and 6; and undoubtedly for many other scientific purposes as well. Thus, maintenance of old data sets is a matter of continuing importance.

The cost of remotely sensed data is a matter of considerable concern among researchers. A major issue for the research community is the increasing cost of data maintenance. The sheer volume of remotely sensed data is increasing rapidly as more platforms are launched and as they provide increasingly finer resolution. The costs of transforming the raw data into useful forms, of cataloguing them, of storing and maintaining them, and of making them available increase more rapidly than the volume of data because data archives must continually maintain the old data as well as the new. The volume of data increases as new data come in, and new technology often brings the additional problem of translating between different forms of data storage. The cost issue is multiplied further as data from military satellites are declassified and become available. In many cases, military and intelligence organizations have no use for the old data and therefore no incentive to make them useful for social science, regardless of their potential value. Yet the older material from military sources may be especially important for social science because it extends time series further backward with levels of spatial resolution that are not available from any other source for historic data. Thus, the cost of data systems is growing in importance just as the data are becoming increasingly useful for social science.

Another cost issue concerns the cost of data to potential users. Here, the situation is highly fluid. U.S. government policy has so far kept remotely sensed images in government data systems relatively inexpensive to scientists. However, cost depends on the platform and on whether any government agency has ordered a particular scene. In addition, budgetary pressures and tendencies toward commercialization of data systems may alter the situation. We believe it is important for science that the government maintain its policy of keeping remotely sensed data inexpensive or make sufficient research funding available so that scientists have access to the necessary data.

PLAN OF THE VOLUME

This volume is intended to be useful to researchers and research sponsors who are considering what they can do to foster or participate in collaborations between remote sensing and social science. It is divided into two main sections. The first, consisting of Chapters 1 through 3, provides conceptual and historical background. The second, consisting of Chapters 4 through 10, offers case examples that illustrate the uses and potential applications of remote sensing for social scientific purposes. Each of these case examples describes a research area in which the effort to link remote sensing and social science shows promise for advancing knowledge. The cases also indicate what is involved, both intellectually and in practical terms, in achieving that promise. The most intensive coverage is given to research on land-use change in rural areas of developing countries (Chapters 4 through 7) because this is currently the area of the most intensive research activity involving collaboration between social scientists and remote sensing specialists. Chapters 8 through 10 present applications to urban land-use issues, famine early warning, and public health that illustrate some promising frontiers for social scientific use of remotely sensed data. The volume does not include other actual or potential social science applications of remote sensing. For example, remote sensing is used in mapping the impacts of and recovery from natural disasters and can be used in research on disaster response. It may also be possible to link remotely sensed data on atmospheric trace gas concentrations to ground-based data on industrial activity in order to improve models that link human activities to their environmental consequences.

In Chapter 2, Morain provides a historical perspective by examining past relationships between remote sensing and social science, focusing especially on the long history of the Landsat program. Although there has recently been a proliferation of remote sensing platforms that have great potential usefulness for social science, Landsat and AVHRR have until now been the main space-based platforms used in social science research. The chapter, though perhaps fragmentary from a remote sensing perspective, provides a good account of the history of the sources of space-based data most commonly combined with social science data. A good source for more detail on the full range of remote sensing technology is Ryerson (1996).

In Chapter 3, Geoghegan and colleagues discuss the major issues that emerge from the most extensive current effort to link remote sensing and social science—the Land-Use/Cover Change (LUCC) research program, sponsored by the International Geosphere-Biosphere Programme and the International Human Dimensions Programme on Global Environmental Change. Among these issues are those of spatially explicit modeling in the social sciences, analyses that make links across spatial scales and levels of analysis, and the problem of developing effective concepts and analytical methods for simultaneously analyzing changes in time and in space.

Chapters 4 and 5 describe research projects in Amazonia that use remote sensing of land cover to examine such issues as the effects of human population dynamics on deforestation and the effects of deforestation on land-use change. The two studies, though focused on the same geographical region, differ greatly in the levels of analysis at which they examine data and in the variables they select for study. Chapter 6 examines population dynamics and land-use and land-cover change at the village level in one district in Thailand. Of particular interest is the project's effort to use both social and remotely sensed data in time series to illuminate mutual causation between population dynamics and land-use change. Chapter 7 examines land-use change in the Petén region of Guatemala. As in other areas where deforestation is progressing, remote sensing is used to track the process and its relationships to social driving forces such as road construction and land development or preservation policies. The chapter, which focuses on an important region for archaeological research, shows how remote observation has been used to identify sites not previously discovered from the ground.

Chapters 8, 9, and 10 present some promising frontier areas for linking remotely sensed data and social science. These areas are promising because remote sensing has already demonstrated its relevance to socially important phenomena; greater integration of social science concepts is likely to yield further practical and scientific advances. In Chapter 8, Cowen and Jensen identify a variety of remotely sensed attributes that could be used for analysis of urban dynamics in the United States. They specify the degree of spatial and temporal resolution required to measure change in these attributes and, in an important contribution, compare these data requirements with the capabilities of existing remote sensing platforms. The comparison suggests some areas in which existing remotely sensed data could be used more extensively in social science research and others in which such uses are likely to become possible as soon as higher-resolution remote data become available. For example, remote platforms might be able to measure an underinvestigated and potentially important set of social variables represented by the physical characteristics of neighborhoods. Arguably, the behavior of significant outsiders, such as social service providers and mortgage lenders, is shaped more by stereotyping based on a neighborhood's observable physical characteristics than by the actual attributes of its inhabitants. Chapter 8 also provides an illustration of the use of remotely sensed data to model and predict the course of urban development.

Hutchinson (Chapter 9) discusses some applications of remote sensing and social science to famine early warning systems in Africa. Remote data are increasingly useful for monitoring and forecasting crop production—the supply side of famine—and also valuable for measuring roads and other infrastructure important to food distribution. However, famine also depends on the economic demand for food and the economic and political institutions that allocate food or restrict its availability—factors much more easily measured by standard social scientific methods. Thus, famine early warning can best be accomplished by collaborative efforts. The same can probably be said for basic understanding of food systems and their implications for the nutritional status of populations. It has often been argued, for example, that famine is more often a result of inadequate food distribution due to war, economic inequality, or the practices of repressive governments than of supply shortfalls. Nevertheless, famine and nutritional deficits are almost certainly the result of an interaction of factors: often, both supply shortages and interferences with distribution are necessary conditions. The study of how food production interacts with various political and economic forces is likely to be advanced by combining remotely sensed and ground-based data within the same analytical schemes.

In Chapter 10, Epstein reports on uses of remote sensing to monitor and anticipate the emergence of infectious disease outbreaks. Until now, this work has involved contributions from remote sensing experts, ecologists, epidemiologists, and public health specialists without explicit social science involvement. However, there are opportunities for the involvement of researchers in such fields as disease prevention, health promotion, and risk communication, as well as for interdisciplinary research linking the ecology of disease organisms and vectors with human ecology.

Two appendices are intended as resources for social scientists who are relatively unfamiliar with remotely sensed data. Appendix A provides a guide to numerous major sources of remotely sensed data. Because the data sets change so rapidly, this appendix will be updated periodically on the World Wide Web. Appendix B offers a glossary of technical terms used both in remote sensing and in social science with which experts in either field might need to be familiar when venturing into the other.

ACKNOWLEDGMENT

We thank John Estes and B.L. Turner II for their very helpful comments on earlier drafts.

NOTES

1 Definitions of this and other technical terms can be found in Appendix B.

2 Availability of satellite data changes rapidly. Appendix A identifies a number of current information sources and a Web site that will provide updated information. At the time of this writing, we are told that several improvements in resolution are just over the horizon. The French Système Pour l'Observation de la Terre (SPOT) satellite, scheduled for launch in 1998, is to include a 20-meter resolution mid-infrared band in the 1.55-1.75 microns range, 10-meter resolution in the red and green visible bands and the near infrared, and 2.5-meter resolution in the panchromatic band. Updated information on this satellite may be obtained from WWW.spot.com. A private company, EarthWatch Inc., launched the EarlyBird satellite in December 1997 with 3-meter resolution and plans to launch the QuickBird satellite in 1998 with 1-meter resolution, both panchromatic. Updated information on these satellites may be obtained at WWW.digitalglobe.com. India plans to launch a series of satellites with 1-meter resolution, IKONOS 1 and 2, in late 1997 and 1998. In addition, declassified high-resolution data from military and intelligence satellites from the 1960s and early 1970s are beginning to be made available.

3 Partly for economy of exposition and partly to make some of our points more emphatically, we refer to social science and remote sensing as fields. One reviewer questioned whether remote sensing is a "field," a point that raises the question of whether social science is a "field." The latter question is the easier of the two: social science is actually a collection of fields. The status of remote sensing is more ambiguous. As the reviewer correctly noted, remote sensing is in one sense only a technology or a tool. Courses in remote sensing are taught in a variety of departments, including geography, geology, landscape architecture, oceanography, and forestry. However, remote sensing also has many of the characteristics of a field. There is a specialized language that those in the remote sensing area know and use, and outsiders do not understand. There are scientific journals devoted to remote sensing topics, and there are annual meetings for those who speak the remote sensing language. Thus, despite the room for disagreement, we refer to both remote sensing and social science as "fields."

4 With few exceptions (e.g., Chapter 8), the researchers represented in this volume have used satellite data rather than aerial photographs. Among the studies we know that use remotely sensed data and social science data together, the vast majority use satellite data rather than aerial photographs. There are probably a number of reasons for this. While the situation varies from country to country and region to region, in general it is more difficult to find the needed images in the form of aerial photographs than as satellite images. Satellites, because they orbit the earth, collect data for much of the earth's surface, while aerial photographs cover much smaller sections of the earth. Further, the various organizations that have emerged to sell and distribute satellite data tend to have regional or global coverage. On the other hand, those that distribute or sell aerial photographs tend to operate at the national level, or even finer scales. In addition, for many areas of the world, the search costs to determine the availability of aerial photographs are formidable. Satellite data cover

more of the spectral range, sometimes including the near and thermal infrared, whereas aerial photographs are typically panchromatic (black and white). Finally, satellite data typically come in digital format allowing for easier incorporation into a geographic information system (GIS), but aerial photographs are typically in analog format, thus requiring additional work before they can be incorporated into a GIS.

5 Roads illustrate some of the limits of remote sensing for measuring visible phenomena. Most sensors "see" the highest level of cover between the satellite and the earth, although there are exceptions, such as radar data. Thus on a cloudy day, most sensors will see the clouds. If a road is tree lined and well shaded by the trees, on a clear day the typical satellite platform will provide an image of the trees and not the road.

6 The specifics, of course, depend on the constellation of techniques being used to locate all households.

7 Most satellite platforms do not capture the poles. However, given the limited human population at the poles, we can consider the data from most satellite platforms to be global from a social science perspective. Even though the data from many satellite platforms are essentially global, the uses of the data need not be global. Indeed, the examples in this volume are much more localized.

8 Aerial photographs are not subject to this pixel constraint. Instead, properties of the photography determine the minimum mapping unit, which can function like a pixel in that it determines what one can "see" in an aerial photograph.

9 Market research firms typically compile social data at much lower levels of aggregation than governments do—in the United States, at the level of the postal zip code or even the zip-plus-four, which typically corresponds to a geographic area that encompasses the residences of a few dozen households. Some elements of these privately held data sets, such as data on consumer expenditures, are parallel to data collected by governments but available to clients at finer resolution. These data have not to our knowledge been used much in social science research.

REFERENCES

Bureau of the Census
   1993   *1990 Census of Population. Social and Economic Characteristics: United States.* Bureau of the Census, U.S. Department of Commerce. Washington, D.C.: U.S. Government Printing Office.
Carls, N.
   1947   *How to Read Aerial Photographs for Census Work.* Washington, D.C.: U.S. Government Printing Office.
Colwell, R.R.
   1996   Global climate and infectious disease: The cholera paradigm. *Science* 274:2025-2031.
Estes, J.E., J.R. Jensen, and D.S. Simonett
   1980   Impacts of Remote Sensing on U.S. Geography. *Remote Sensing of Environment* 10:43-80.
Faust, K., B. Entwisle, R.R. Rindfuss, S.J. Walsh, and Y. Sawangdee
   1997   Spatial Arrangement of Social and Economic Networks among Villages in Nang Rong, Thailand. Paper presented at the annual meeting of the Sunbelt Social Network Conference, San Diego, Calif.
Hosenball, S.N.
   1990   International and U.S. domestic law governing remote sensing. Pp. 125-140 in *Earth Observation Systems: Legal Considerations for the '90s.* Bethesda, Md. and Chicago, Ill.: American Society for Photogrammetry and Remote Sensing and American Bar Association.

Lucas, R.M., M. Honzak, G.M. Foody, P.J. Curran, and C. Corves
  1993   Characterizing tropical secondary forests using multi-temporal Landsat sensor imagery. *International Journal of Remote Sensing* 14:3061-3067.
Moran, E., E. Brondizio, P. Mausel, and Y. Wu
  1994   Integrating Amazônian vegetation, land use, and satellite data. *BioScience* 44(5):329-338.
National Research Council
  1997   *Rediscovering Geography: New Relevance for Science and Society.* Rediscovering Geography Committee. Washington, D.C.: National Academy Press.
Rindfuss, R.R., S.J. Walsh, and B. Entwisle
  1996   Land Use, Competition, and Migration. Paper presented at the annual meeting of the Population Association of America, New Orleans, La.
Ryerson, R.A., ed.
  1996   *Manual of Remote Sensing.* 3rd edition (CD-ROM). Bethesda, Md.: American Society for Photogrammetry and Remote Sensing.
Skole, D.L., W.H. Chomentowski, W.A. Salas, and A.D. Nobre
  1994   Physical and human dimensions of deforestation in Amazonia. *BioScience* 44(5):314-322.
Turner II, B.L.
  in   Frontiers of Exploration: Remote Sensing and Social Science Research. In *Proceedings*
  press   *of Pecora 13.* Bethesda, Md.: American Society for Photogrammetry and Remote Sensing.
Uhlir, P.F
  1990   Applications of remote sensing information in law: An overview. Pp. 8-22 in *Earth Observation Systems: Legal Considerations for the '90s.* Bethesda, Md. and Chicago, Ill.: American Society for Photogrammetry and Remote Sensing and American Bar Association.

2

A Brief History of Remote Sensing Applications, with Emphasis on Landsat

Stanley A. Morain[1]

By 1997, at least six civilian land remote sensing satellite systems were being operated by the United States, France, India, Japan, Canada, Russia, and the European Space Agency. On command, all of them make measurements of the land surface, transmitting spectral data to a global network of strategically located ground receiving stations. Data from these earth-observing satellites are used to map, monitor, and manage the earth's natural and cultural resources.

Perhaps the first person who believed that not only machines but humans, too, could venture into space was Jules Verne, a French provincial lawyer with no scientific or technical training (Mark, 1984). Verne made the extraordinary prediction that a rocket would be launched from Florida by means of chemical propulsion, and that the crew would include three people (and a dog). First they would only circle the moon and return to earth, as did Apollo 8. This would be followed by a trip to the moon's surface, the return to earth ending with a "splash down" in the Pacific Ocean and recovery by a warship. Perhaps Verne's most remarkable prediction was that Americans would make this first journey. What he did not predict was that astronauts would be awed by the blue marble, or that their photographs would so sensitize the world that subsequent human scientific interest would shift toward space as a means for studying the earth. The United States was not only the first to land on the moon, but beyond Verne's vision, it also developed the first remote sensing satellites, whose profound importance to today's concept of a global village cannot be overstated.

Most histories of remote sensing identify Gaspard Felix Tournachon as the first person to photograph "remotely," using balloons above Paris in 1859. Balloons were also used for aerial reconnaissance during the American Civil War.

By 1909, aerial photographs were being taken from airplanes for a wide range of uses, including warfare, land-use inventory, and publicity. Aerial photography remains an important application of remote sensing, with a sophisticated range of cameras being used to collect information on geology, land use, agricultural conditions, forestry, water pollution, natural disasters, urban planning, wildlife management, and environmental impact assessments (Lillesand and Kiefer, 1987). Evelyn Pruitt of the Office of Naval Research originally coined the term "remote sensing."

GROWTH OF THE REMOTE SENSING COMMUNITY

The Environmental Research Institute of Michigan (ERIM) is credited with organizing the first technical conference on remote sensing in the United States, perhaps in the world. Its First Symposium on Remote Sensing of Environment, sponsored by the Navy's Office of Naval Research (ONR), had 15 presenters and 71 participants (Environmental Research Institute of Michigan, 1962). In the audience was Dr. William Fischer of the U.S. Geological Survey (USGS), an early advocate of an earth-observing system. The field was so new that Dana Parker's inaugural address focused on fundamentals of the electromagnetic spectrum. In October 1962, the second symposium was held, drawing 162 participants to hear 35 technical papers (Environmental Research Institute of Michigan, 1963). It was sponsored by the Geography Branch of ONR, the Air Force Cambridge Research Laboratory, and the Army Research Office.

At the third symposium in October 1964, 280 participants heard 54 technical papers (Environmental Research Institute of Michigan, 1965). By this time, all of the principal government, academic, and private-sector motivators for an orbiting resource satellite system were represented. Among the papers in the proceedings was one by Dr. Robert Alexander from ONR. He gave the first announcement for what evolved into Landsat-1. His abstract read:

> The National Aeronautics and Space Administration is sponsoring a study of the geographic potential of observations and experiments which might be carried out from the remote vantage of earth-orbiting spacecraft. The investigation will involve both the value of the science of geography and the expected practical applications of an earth-viewing-orbiting laboratory and other possible geographic satellite systems. Early emphasis will be on problems of systematizing and managing the flow of geographic information which would result from such a program. (p. 453)

The eighth symposium (Environmental Research Institute of Michigan, 1972), held 8 years after Alexander's announcement and only a few weeks after the first Landsat-1 images had been released, included 14 presentations describing the utility and quality of the Landsat data. By that time, the broader field of remote sensing had attracted over 700 participants who selected from a program of 116 papers on topics including theoretical and applied engineering, natural and cultural resource monitoring, state and local government applications, and even environmental and public health issues.

The National Aeronautics and Space Administration (NASA) became an official sponsor of ERIM symposia in 1971. In 1973, NASA's administrator inaugurated the agency's decade-long program of University Research Grants to stimulate cooperative research at the local level. In some cases it supported construction of laboratory facilities and supplied the equipment needed to train the 1970s generation of Ph.D. remote sensing specialists. By the mid- to late 1970s, many of these young professionals were employed on collaborative federal government research projects for proof-of-concept applications embracing the whole range of natural and cultural resources. The Large Area Crop Inventory Experiment (LACIE) and its spin-off AgriSTARS are examples of these projects. The Application System Verification Tests (ASVTs) are another.

While these were not the only application development programs under way, they were symptomatic of a massive spontaneous adaptation to fundamentally new ways of studying the earth. Within little more than a decade of ERIM's first symposium, the core remote sensing community had increased its numbers by several orders of magnitude. This community brought about major changes in organizational structures, became the basis for a new international research agenda, and germinated the first seeds of thought on global habitability.

STIMULI FOR AN ORBITING RESOURCES SATELLITE

In 1946, the United States Army Air Corps requested that the RAND Corporation consider how objects might be inserted into orbit (Mark, 1988). The study resulted in a report titled *Preliminary Design of an Experimental World-Circling Spaceship* (Burrows, 1986). The proposed midget moon, or "satellite," would provide "an observation aircraft [sic] which cannot be brought down by an enemy who has not mastered similar techniques." After many aborted liftoffs and system failures, the military successfully launched its first earth-observing satellite in August 1960. It was called Discoverer and was expected to be an unclassified system to support biomedical research and earth observations (Tsipis, 1987; Whelan, 1985). A few months after launch, however, a presidential directive classified the Discoverer program and plunged it into deep secrecy. Only recently have images collected by its successor, the Corona program, been declassified for public use (McDonald, 1995).

In parallel with the early military/intelligence programs in space, the scientific and industrial communities in the United States were awakening to the potential of space for providing a new world perspective. In 1951, 6 years before Sputnik 1, Arthur C. Clarke, a science fiction writer and prophet of technology, proposed that a satellite could be inserted into orbit over the north and south poles while the earth revolved beneath it, allowing humans to view the planet in its entirety (Fink, 1980). In April 1960, NASA and the Department of Defense

(DOD) launched the Television and Infrared Observational Satellite (TIROS-1) into such an orbit, inaugurating the first experimental weather satellite (U.S. Department of Commerce and National Aeronautics and Space Administration, 1987). This system generated the first television-like pictures of the entire globe in a systematic and repetitive manner. The ongoing series of TIROS satellites became operational in 1966 as the TIROS Operational Satellites (TOS), renamed the National Oceanic and Atmospheric Administration (NOAA) Polar Orbiting Environmental Satellites (POES) in 1970 (Morain and Budge, 1996).

Although the early weather satellites could not provide details on land processes or use, they were invaluable in understanding the weather and thus providing warnings of weather-related natural disasters, as well as information on rainfall of relevance to agriculture and water management (Lillesand and Kiefer, 1987).

The series of NOAA weather satellites carrying the Advanced Very High Resolution Radiometer (AVHRR) from 1978 on was important not only to meteorological research and prediction, but also to studies of vegetation and land use. Over the last two decades, AVHRR data have been used to construct vegetation indices for monitoring crop failures, urban climate, locust outbreaks, range conditions, deforestation, and desertification (Ehrlich et al., 1994; Gallo et al., 1993; Townshend and Tucker, 1984; Tucker et al., 1991). The NOAA satellites have been important in the development of famine warning systems, such as those described by Hutchinson in this volume. These satellites are of much coarser resolution than the Landsat series discussed below (one AVHRR image can cover as many as 100 Landsat images), but the lack of detail is somewhat compensated by the broad coverage, lower costs, and more frequent (twice daily) flyovers. Roller and Colwell (1986) argue that the AVHRR images can be used very efficiently to stratify land use and topography for more detailed studies with Landsat and other higher-resolution satellites. At an even coarser scale, the Geostationary Earth-Orbiting Satellite (GOES) provides the continuous hemispheric coverage of cloud cover and other aspects of the atmospheric circulation shown on evening weather forecasts as a visual confirmation of approaching weather, particularly extreme events such as hurricanes. Other satellites, launched predominantly for ocean research applications, such as Nimbus, SeaSat, and SeaWifs, have provided information of relevance in coastal zone studies and fisheries management (Consortium for International Earth Science Information Network, 1997).

The photographs taken from manned space expeditions such as Gemini and Apollo were used in several land inventory applications. In 1973, Skylab took more than 35,000 images that have become classics in many resource management and earth science texts. In recent years the Space Shuttle has taken numerous images of sites of human interest, many at the request of researchers concerned with deforestation, urbanization, pollution, and water resource management.

The United States pioneered land remote sensing from space and has been the unquestioned leader in the development and application of this unique technology. Americans take pride in having developed the Landsat program, as well as other, more recent civilian programs. The evolution of Landsat, however, has been neither linear nor predictable. The remainder of this chapter provides an overview of its conception, genesis, and growth; its accomplishments and current status; and its uncertain future. At this point, Landsat still dominates remote sensing applications in the United States.

FORCES MOTIVATING THE DEVELOPMENT
OF REMOTE SENSING

At first, civilian space remote sensing consisted of experimental missions to develop proofs of concept for any applications that had a sound scientific basis—and some, perhaps, that did not. All of the missions conducted in the 1960s and 1970s, including those that acquired hand-held Gemini, Apollo, Skylab, and Apollo-Soyuz photography and the early Landsats, were approved and funded as experiments to advance space science. The forces that stimulated and motivated today's Landsat program were numerous and complex. Five of the most compelling were (1) the need for better information about the earth's features, (2) national security, (3) commercial opportunities, (4) international cooperation, and (5) international law.

Need for Better Information

Society requires better information about the geographic distribution of the earth's resources, and satellites help in obtaining this information. Earth now supports more than 5 billion people, and human populations are growing at 1.5 percent per year, or 3 people per second. By 2000, the world's population will exceed 6 billion. Nobody knows how many people the earth can sustain; some guess 8 billion, while others say nearly double that number (McRae, 1990; Ashford and Noble, 1996). Regardless of how many people can be squeezed onto the planet, however, there are limits to the renewable and nonrenewable resources needed to support them. Efficient management of renewable resources and judicious use of nonrenewable resources, as well as improved conservation and protection of fragile and endangered environments, depend on timely information about, and accurate analysis of, those resources. In the late 1960s, there was a convergence of thought that the best means for acquiring the needed data was earth-orbiting satellites that could provide continuous and nearly synoptic coverage of terrestrial resources. This would be the case in particular for understanding and measuring earth system processes at regional, continental, or global scales. Human numbers and human impacts on resources thus became an early and globally compelling argument for studying the earth from space.

National Security

The U.S. government maintains national security, a mission that includes using data from civilian satellites to protect and defend the nation against aggressors. It is no secret that defense/intelligence satellites are assets for maintaining national security. It is not as widely known, however, that the defense/intelligence community has always used data from civilian satellite systems in carrying out its security mission (National Space Council, 1989). While there were, and still are, many security limitations imposed on the first generation of earth-observing systems, there was nevertheless a defensible argument that such a system should be developed. It was recognized that timely information about the global distribution of critical natural resources and the factors that affect global environmental conditions is integral to national security and would be gleaned in part from civilian systems. Indeed, the decision to build and launch Landsat-7 was driven partly by requirements of the defense/intelligence community (Office of the President, 1992).

Commercial Opportunities

The U.S. government encourages private-sector investment in the nation's space program, including civilian earth-observing satellites. Remote sensing technology was developed by aerospace industries under contract to federal government agencies to satisfy both government and public needs. Commercialization of this know-how is fundamental to American ideals and has been a stimulus for continued industry investment. By the early 1970s, several industries, including communications satellites and booster launch services, had already proven the commercial value of the space environment. The prospects for similar financial gain from data on the earth's resources seemed self-evident, but a successful experimental system would be a necessary first step. The assumption that data on the earth's resources would have commercial value beyond their benefit for the public good was thus a powerful argument for developing the Landsat program. Full commercialization of both the space and ground segments may yet prove to be intractable, but there is clearly a viable and profitable role for industry in building space platforms, sensor systems, and ground processing facilities, as well as providing value-added data processing services. The commercial value of space-based remote sensing products and services is a hypothesis that will finally be tested with several privately owned satellites scheduled for launch in 1998.

International Cooperation

The U.S. government seeks international cooperation on civilian earth-observing satellites in order to better understand, manage, share, and protect the earth's resources. The United States is committed to using space for peaceful and defense purposes only. To this end, Americans want to share benefits from space technology with other nations, but they also want to protect their commercial interests. Earth observations from space have never been the sole domain of the United States, and several nations now participate in this activity with competing spacecraft and sensor systems. The argument for promoting cooperation among nations was originally driven by opportunities for America to promote its foreign policy objectives, but have evolved to include elimination of unnecessary redundancies among different national programs and the savings that might be realized from joint programs through cost sharing (Office of the President, 1996). While these objectives were not publicly articulated in the early 1970s, they were a driving political force in the Landsat planning process.

International Law

Societies are governed by laws, rules, and regulations to maintain organization and order—not only on earth, but also in space. Societies establish laws by which they govern against chaos and anarchy. Space law is relatively new to jurisprudence, but it is a central force because it sets the rules by which all nations, not just the space-faring ones, have a voice in how to participate in space technology. Legal aspects of civilian space-based remote sensing are complicated and sometimes controversial, especially regarding the issues of national sovereignty, rights of privacy, and, most recently, commercial gain. The United States has always argued strongly for an open-skies and nondiscriminatory data distribution policy for civilian space data, believing the greatest good for the greatest number will come from free and open exchanges of data and information (Stowe, 1976; Office of the President, 1988). When the United States implemented the Landsat program, it made an extraordinary effort to ensure that every nation had access to these data, even to the extent that foreign ground receiving stations were installed.

THE EVOLUTION OF LANDSAT

The Landsat Concept

The concept of a dedicated, unmanned land-observing satellite emerged in the mid-1960s from the complex milieu of synergisms and conflicting interests described above. It arose primarily in ONR and USGS under the latter's late director Dr. William T. Pecora (Waldrop, 1982). In fact, scientists within USGS, working in cooperation with Dr. Archibald Park and others in the U.S. Department of Agriculture (USDA), originally proposed to the (then) Bureau of Budget (now Office of Management and Budget) that an Earth Resources Observation Satellite (EROS) be built, launched, and operated. The Under Secretary of the Interior announced the objectives of EROS in a memorandum dated July 12,

1967, and addressed to the Department of the Interior's assistant secretaries and bureau heads (Luce, 1967). These objectives were to (1) construct and fly an earth-observing system by the end of 1969, and follow this with improved and modified systems as required by the operational needs of resources programs; (2) provide unclassified remotely sensed data to facilitate the assessment of land and water resources of the United States and other nations; and (3) design specific systems on the basis of users' data requirements, distribute such data to users, and make operational use of the data in resource studies and planning. The overall goal of the proposed EROS program was to acquire satellite remotely sensed data in the simplest possible way, deliver these data to users in an uncomplicated form, and ensure the data's easy use (Pecora, 1972).

Since development of space technology was the responsibility of NASA, the Department of the Interior's proposal was rejected. NASA Administrator James Webb met with President Johnson to discuss Interior's announcement, and Webb succeeded in retaining control of what was to become an "experimental" program (Covert, 1989). In cooperation with the Department of the Interior, USDA, and other agencies, NASA designed an earth-observing satellite, obtained funding for the project, and successfully launched the first Earth Resource Technology Satellite (ERTS-1) in July 1972.

Although unsuccessful with its own satellite system, the Department of the Interior continued with an EROS program under the direction of USGS. The EROS mission was to archive and distribute remotely sensed data, and to support remote sensing research and applications development within Interior.[2] To carry out the EROS responsibilities, USGS built the EROS Data Center (EDC) in 1972.

After the launch of Landsat-1, Goddard Space Flight Center (GSFC) hosted three symposia in quick succession (National Aeronautics and Space Administration, 1973a, b). These symposia were designed especially for their Landsat-sponsored investigators to report "user identified significant results." The application categories were agriculture/forestry, environment, geology, land use/land cover, water, and marine. Each of the proceedings approached 2000 pages of text and graphics, mainly detailing early application concepts and models. The Landsat program had such a powerful impact in so many arenas that it was declared operational in late 1979 after a prolonged debate among participating government agencies (U.S. Department of Commerce, 1980).

In the decades to follow, however, Landsat-1 replacement satellites were the subject of severe political uncertainty. The program witnessed a change of guard among its staunchest supporters, and the satellites were casually labeled a "technology in search of an application." Kuhn's (1962) prescription for scientific revolutions foreshadowed these developments by predicting a period of scientific uncertainty, if not outright denial, by whole sectors of the science and technology community. In the waning years of the twentieth century, complexity science views such developments as inherent to self-adaptation in complex systems (Waldrop, 1992). Once a critical mass of support had been attained, the individual actions of sensor developers, data suppliers, data analysts, and end-users ensured the continuity of the technology, even if its development seemed chaotic, and even if the direction of that development was unclear. After a quarter-century of successful data gathering, the fate of the Landsat program remains uncertain, but the technology derived from the program continues to permeate user communities and become more complex as the applications it has spawned mature.

Even as the first Landsat was being prepared for launch, conflicts in agency roles had begun to appear. NASA's charter was to engage in space research and technology development. That charter did not include earth resource data handling, processing, archiving, or distribution to a large and diverse scientific community, or to an even larger group of public and private users. Consequently, NASA reached agreement with several resource management agencies to transfer responsibility for the program's ground segment, while NASA retained responsibility for the space segment.

The Landsat System

ERTS-1 was launched from Vandenberg Air Force Base in California. A Nimbus-type platform was modified to carry the sensor package and the data relay equipment. ERTS-2 was launched on January 22, 1975. It was renamed Landsat-2 by NASA, which also renamed ERTS-1 as Landsat-1. Three additional Landsats were launched in 1978, 1982, and 1984 (Landsat-3, -4, and -5, respectively). As documented by USGS (1979) and by USGS and NOAA (1984), each successive satellite system had improved sensor and communications capabilities (see Table 2-1).

TABLE 2-1 Background Information and Status of Landsat Satellites

Satellite	Launched	Decommissioned	Sensors
Landsat-1	July 23, 1972	January 6, 1978	MSS and RBV
Landsat-2	January 22, 1975	February 25, 1982	MSS and RBV
Landsat-3	March 5, 1978	March 31, 1983	MSS and RBV
Landsat-4	July 16, 1982	[a]	TM and MSS
Landsat-5	March 1, 1984	[b]	TM and MSS
Landsat-6	October 5, 1993	[c]	ETM
Landsat-7	May 1998	[d]	ETM+

[a]In standby mode
[b]Operational
[c]Never achieved orbit
[d]Anticipated launch

Landsats-1, -2, and -3

The first three Landsats operated in near-polar orbits from altitudes of 920 km. They circled the earth every 103 minutes, completing 14 orbits a day and producing a continuous swath of imagery 185 km wide. To provide nearly complete coverage of the earth's surface, 18 days and 251 overlapping orbits were required. The amount of swath sidelap varied from 14 percent at the equator to nearly 85 percent at latitude 81° north or south. These satellites carried two sensors: a Return Beam Vidicon (RBV) and a MultiSpectral Scanner (MSS). The RBV sensor was a television camera designed for cartographic applications, while the MSS was designed for spectral analysis of terrestrial features. The MSS sensor scanned the earth's surface from west to east as the satellite moved in its descending (north-to-south) orbit over the sunlit side of the earth. Six detectors for each spectral band provided six scan lines on each active scan. The combination of scanning geometry, satellite orbit, and earth rotation produced the global coverage originally suggested by Arthur C. Clarke for viewing the earth's entire land surface. The spatial resolution of the MSS was approximately 80 m with spectral coverage in four bands from visible green to near-infrared (IR) wavelengths (see Table 2-2). Only the MSS sensor on Landsat-3 had a fifth band, in the thermal-IR. ERTS-1 delivered high-quality data for almost 4 years beyond its designed life expectancy of 1 year and was finally shut down on January 6, 1978. Landsats-2 and -3 were decommissioned in February 1982 and March 1983, respectively.

TABLE 2-2 Radiometric Range of Bands and Resolution for the MSS Sensor

Band		Wavelength	Resolution
Landsats -1,-2,-3	Landsats -4,-5	(micrometers)	(meters)
4	1	0.5-0.6	79/82[a]
5	2	0.6-0.7	79/82
6	3	0.7-0.8	79/82
7	4	0.8-1.1	79/82
8[b]		10.4-12.6	237

[a]The nominal altitude was changed from 920 km for Landsats-1 to -3 to 705 km for Landsats-4 and -5, which resulted in a resolution of approximately 79 and 82 m, respectively.
[b]Landsat-3 only.

SOURCE: U.S. Geological Survey and National Oceanic and Atmospheric Administration, 1984.

Landsats-4 and -5

Landsats-4 and -5 carried both the MSS and a more advanced sensor called the Thematic Mapper (TM).[3] Their orbits were somewhat lower than those of their predecessors at 705 km and provided a 16-day, 233-orbit repeat cycle with image sidelap that varied from 7 percent at the equator to nearly 84 percent at latitude 81° north or south. The MSS sensors aboard Landsats-4 and -5 were identical to earlier ones. Both sensors detected reflected radiation in the visible and near -IR bands, but the TM sensor provided seven spectral channels of data as compared with only four channels collected by MSS. The wavelength range for the TM sensor spanned the blue through mid-IR bands (see Table 2-3). The 16 detectors for the visible and mid-IR bands in the TM sensor provided 16 scan lines on each active scan. The TM sensor had a spatial resolution of 30 m for the visible, near-IR, and mid-IR bands and a spatial resolution of 120 meters for the thermal-IR band. As with all earlier Landsats, sensors on these satellites imaged a 185-km swath. Today, Landsat-4 has lost all capability to communicate TM data and is in standby mode. Landsat-5 has lost its Tracking and Data Relay Satellite System (TDRSS) capability, but continues to provide data via direct downlink to the United States and the international ground stations.

TABLE 2-3 Radiometric Range of Bands and Resolution for the TM Sensor

Band, Landsats -4,-5	Wavelength (micrometers)	Resolution (meters)
1	0.45-0.52	30
2	0.52-0.60	30
3	0.63-0.69	30
4	0.76-0.90	30
5	1.55-1.75	30
6	10.40-12.50	120
7	2.08-2.35	30

SOURCE: U.S. Geological Survey and National Oceanic and Atmospheric Administration, 1984.

Landsat-6

Landsat-6 was launched on October 5, 1993, but failed to achieve orbit. It was similar to Landsats-4 and -5 in terms of spacecraft design and planned orbital configuration. The MSS and TM sensors were replaced by an improved TM sensor called the Enhanced Thematic Mapper (ETM), from which no data were received.

ASSESSING THE IMPACT

Beyond the personalities and visions that led to Landsat-1, the program's course could not have been charted or predicted; it had to be experienced. Landsat-1 not only inaugurated a global research agenda, but also spawned a genre of careers in engineering and the natural sciences. Arguably, Landsat-1 provided academic geographers with real-world data for applying and testing their theoretical models, thus giving their discipline access to its first new set of spatial analytical tools since the electronic calculator. Landsat-1 at first augmented and then gradually changed the 1960s approach to remote sensing as a multispectral tool, making it possible to add time to the analytical tool kit for the earth's resources.

As expected, Landsat-1 promoted business applications for data on the earth's resources and stimulated a proliferation of complementary international platforms. Both the American and International Societies for Photogrammetry quickly added Remote Sensing to their organizational titles as adoption of the technology produced a dramatic increase in new members and research foci. In short, Landsat-1 broadened participation and coalesced a disparate community of practitioners into an international body whose collective efforts produced a new remote sensing paradigm. As with all such emerging phenomena, the growth of remote sensing technology was partly ordered and partly chaotic; after July 23, 1972, the community self-organized into a complex system of technology developers, data suppliers, and data analysts/users. Landsat-1 data became the keystone around which the technology would adjust and grow.

A New Paradigm

A basic premise of remote sensing is that the earth's features and landscapes can be discriminated, identified, categorized, and mapped on the basis of their spectral reflectances and emissions. Pre-Landsat literature from the ERIM symposia reveals this focus. At that time, sensor designs spanned the electromagnetic spectrum from ultraviolet wavelengths to passive and active microwave frequencies. The multispectral concept combined sensors across these electromagnetic regions, and partitioned within them, to study the spectral domains of the hydrosphere, lithosphere, biosphere, and atmosphere. NASA, among other government agencies, contracted with industry to develop 12-, 24-, and 48-channel scanners for aircraft research in geology, agriculture, forestry, and land use/land cover. Major emphasis was on building libraries of spectral reflectances under controlled laboratory conditions and through data gathered by aircraft. Interpretation keys and crude algorithms for machine processing were commonly employed to identify features, but with a persistent apprehension that such results were limited to that study area's specific time and space.

The Landsat-1 MSS sensor fit into this paradigm by being a four-channel, wide-bandwidth scanning system designed to provide first-order observations of surface covers from space altitudes for essentially all of the earth's terrestrial surface. These basic phenomena included the global land/water interface, vegetated/unvegetated areas, forested/unforested lands, urban/nonurban areas, and agricultural/nonagricultural lands. Each of these categories served as the foundation for formulating upward-spiraling interpretations of human economic uses of the land, for assessing environmental health, and for addressing what would later be called earth system science (U.S. National Aeronautics and Space Administration, 1988).

It was, moreover, recognized that, by virtue of its 18-day orbital repeat cycle, Landsat-1 would offer scientists their first unrestricted opportunity to observe synoptic changes in surface covers that would be impossible to obtain using aerial platforms. The temporal dimension of remote sensing had always been appreciated, but seldom employed usefully outside DOD because high-quality time-series data were essentially nonexistent. With Landsat-1, the time dimension not only was a key design parameter, but also was immediately recognized by the scientific community as an essential ingredient in spectral analyses. By holding solar azimuth relatively constant with an equatorial crossing of approximately 9:30 a.m., the orbital design offered an opportunity to calibrate spectral readings radiometrically across latitudes and longitudes and throughout the annual greening and yellowing cycles of vegetation. Attention shifted sharply away from building spectral libraries to monitoring temporal changes and patterns.

Time was also the enabling parameter for promoting a deeper understanding of physical models in several land analysis applications (Reeves, 1975; Colwell, 1983). In surface hydrology, for example, measurements from data collection platforms were merged experimentally with Landsat-1 data to monitor spatial and temporal changes in the water levels of Lake Okeechobee in order to optimize swamp ecology and balance Miami's urban water needs. Run-off prediction models were augmented by monitoring the geographic extent and depth of river basin snow levels, and temporal dynamics of major floods such as those that occurred along the Mississippi River and Cooper's Creek (Australia) in 1973 were examined in the context of disaster assessment.

Other time-sensitive applications were also advanced. In agriculture, MSS imagery was used to improve an existing production estimation model for wheat in western Kansas, thus proving the concept that satellite-acquired data could provide accurate and timely crop predictions (Morain and Williams, 1975). Forest clear-cuts in Oregon and Washington were monitored, and in Washington, remote observations were actually used to assess lessee compliance. Rangeland studies included spectral responses through time for the assessment of biomass production and general range condition.

Landsat MSS and TM data have also become important in global and local studies of biodiversity and biogeography, and have become key to the emerging framework of "landscape ecology," whereby remote sensing is used to identify vegetation gaps and patterns in the landscape that influence habitats and ecosystem functioning and dynamics.

These early modeling efforts evolved into satellite applications that address today's social and environmental issues (e.g., food security, deforestation impacts, desertification trends, resource sustainability, and news gathering). Yet none of them led directly to these more profound applications. They all required iterations that included many false starts. Early applications, therefore, were important as pioneering efforts and for what they taught the scientific community about future satellite requirements and collateral inputs for problem solving. All of the Landsat-1 results relied on collateral, ground-based data (today's relational database or geographic information system [GIS] technology) and suffered from gaps in temporal data that would have made them more robust. Further, the spectral data were often too coarse. If satellite earth observations were to deliver on their early promise, more spectral channels with narrower bandwidths would have to be acquired from a larger number of platforms providing more frequent observations. It was believed that if this could be achieved, the data and imagery would have commercial as well as public value.

Privatization/Commercialization

As the nation's civilian space research and development agency, NASA successfully executed its role by launching Landsat-1. The handoff of responsibility for data dissemination from NASA to USGS/EDC had already been completed by the time Landsat-1 was launched. The plan was for EDC to serve as the supplier of Landsat products, while NASA would continue to develop future sensors and platforms. Differing agency responsibilities and management agendas, however, plagued the Landsat program from its inception. To resolve these issues, the Carter Administration undertook an extensive review of both the military and civilian space policies. By 1979, new policies had been formulated by which the civilian program was to be made operational, administered by NOAA, and eventually privatized. At about this same time, Congress merged land-, ocean-, and weather-sensing systems under the administration of NOAA.

A crisis ensued (National Research Council, 1985). Among the major players in this crisis were an ever-growing community of Landsat data users, including the news-gathering media, which wanted inexpensive, publicly accessible data; an increasingly vociferous industrial sector that was concerned about pending international competition and believed privatization would preserve America's niche in commercial earth observations; and a federal establishment disinclined to privatize all land, ocean, and weather satellite data systems.

In its effort to reduce the size of government, the first Reagan Administration moved quickly to privatize the Landsat program. What resulted was the Land Remote-Sensing Commercialization Act of 1984 (P.L. 98-965). NOAA solicited bids to manage the existing Landsats and civilian meteorological satellites and, aided by large government subsidies, to build and operate future systems. Bids were received from aerospace companies, an insurance company in New York, a small geoscience firm in Michigan, and a farmer in North Dakota (U.S.

Department of Commerce, 1984). In 1985, a contract was signed with EOSAT Corporation (now Space Imaging/EOSAT Corporation), and the transfer was complete (U.S. Department of Commerce, 1985).

A history of the national debate leading up to and following privatization is provided by Morain and Thome (1990). It is interesting that the most compelling arguments made to Congress for Landsat privatization focused on data and program continuity—not spectral analyses and fine-resolution time-sequential data. Although data continuity has never been defined, and program continuity remains a political question, Congress continues to legislate most aspects of America's space remote sensing activities.

Following another series of program reviews, the National Space Council released its National Space Policy Directive #5, which established new goals and implementation guidelines for the Landsat program (Office of the President, 1992). The directive called for a joint DOD/NASA effort to build, launch, and operate Landsat-7. In October 1992, the Land Remote Sensing Policy Act (P.L. 102-55) was signed into law. It reversed the 1984 decision to commercialize the Landsat system and recognized the scientific, national security, economic, and social utility of land remote sensing from space (Scheffner, 1994). It mandated that DOD and NASA (1) establish a management plan, (2) develop and implement an advisory process, (3) procure Landsat-7, (4) negotiate with EOSAT for a new data policy regarding existing systems, (5) assume program responsibility from the Department of Commerce, (6) conduct a technology demonstration program, and (7) assess options for a successor system.

Scarcely a year had passed before the Landsat program was evaluated for a third time, principally because of severe budget constraints surrounding the High Resolution Multispectral Stereo Imager instrument proposed by DOD for Landsat-7. The National Science and Technology Council (NSTC) recommended that Landsat-7 be developed only with an improved TM instrument and that a new management structure be established so DOD could withdraw from the program. The result was Presidential Decision Directive/NSTC-3, dated May 5, 1994, which reconfirmed the administration's support for the program, but gave NASA, NOAA, and USGS joint management responsibility (Office of the President, 1994). These three agencies negotiated with EOSAT for new Landsat-4 and -5 product prices for the U.S. government and its affiliated users, and are proceeding to develop Landsat-7. Meanwhile a tired but operable Landsat-5 (into its fourteenth year) remains aloft, transmitting consistent and reliable images of the earth to the U.S. ground station and its foreign counterparts.

It has been argued that government policies designed to transfer the Landsat program from the public to the private sector were seriously flawed. These policies did not result in market growth, were more costly to the federal government than would have been the case if the system had been federally operated, did not significantly reduce operating costs, and significantly inhibited applications of the data (Lauer, 1990). Costs of imagery increased from about $200 for an

MSS scene in 1972 to more than $4,500 for a TM scene in 1997, a rate much greater than inflation and prohibitive to many users. Nevertheless, the program continued to provide a flow of high-quality, well-calibrated, synoptic imagery of the earth.

Whether or not Landsat privatization was premature given existing and anticipated markets, it can be argued that a global groundswell of government and academic users, particularly within developing nations in Africa, Latin America, and Asia, stimulated a proliferation of international Landsat look-alike satellites. After 1986, these systems augmented Landsat data around the world, further verifying proof-of-concept applications and elevating overall space-based capabilities to a new plane.

The proposed series of NASA's Earth Observing System (EOS) satellites is designed to provide information on a wider range of variables and at a more detailed resolution than that provided by Landsat, and a number of other nations have launched satellites to provide information of relevance to land-use and other applications. For example, the French Système pour l'Observation de la Terre (SPOT) satellite has a higher resolution (10 m) than that of the latest Landsat satellites and has been used in a variety of applications, including observation of the Chernobyl accident and agricultural monitoring (Sadowski and Covington, 1987). The primary constraint on widespread use of SPOT data is the cost, especially for studies over large areas. India has launched two satellites with sensor systems similar to the Landsat TM, which are being used for natural resources management. Japan and Europe have launched Earth Resources Satellites, which use synthetic aperture radar to provide information on the physical and electrical conditions of terrain. These radar satellites are beginning to provide information of relevance to studies of fire, deforestation, crop monitoring, and urbanization (Consortium for International Earth Science Information Network, 1997; Office of Technology Assessment, 1993).

The Legacy

Landsat-7

Landsat-7 is scheduled for launch in mid-1998. Its payload will be an Enhanced Thematic Mapper (Plus) instrument designated the ETM+. The ETM+ has the same basic design as the TM sensors on Landsats-4 and -5, but includes some conservative advances (Obenschain et al., 1996). It will provide 60-m (as opposed to 120-m) spatial resolution for the thermal band and a full-aperture calibration panel that will result in improved absolute radiometric calibration (5 percent or better). The geodetic accuracy of systematically corrected ETM+ data should be comparable to that characterizing Landsat-4 and -5 TM data, with a specific uncertainty of 250 m (1 sigma) or better. Other features have been added to the Landsat-7 program to facilitate use of the data, particularly by private industry. For example, Landsat-7 will directly downlink ETM+ data to domestic and international ground receiving stations at 150 megabits per second using three steerable X-band antennae. Although transmissions to international ground stations will continue, the system is being designed so that the United States can capture and refresh a global archive to be located at EDC. To make it possible for ETM+ to capture data over regions beyond the range of EDC's receiving antenna, Landsat-7 will use a 378 gigabits per second solid-state recorder capable of storing approximately 40 minutes or 100 scenes of ETM+ data. A second North American receiving station is being added near Fairbanks, Alaska, so that 250 scenes of data per day can be collected. Thus, the recorder will downlink recorded data when the satellite is within range of either EDC or the Alaskan station, and EDC will receive and archive 250 ETM+ scenes per day. These features will provide the capability for global coverage of continental surfaces on a seasonal basis.

On the other hand, Landsat-7 does not include some useful technologies, such as the ability to "point," and its 15 × 15-m spatial resolution may be rapidly overtaken by the 1 × 1-m resolution of the Space Imaging EOSAT system in 1998.

Beyond Landsat-7

The 1992 Land Remote Sensing Policy Act called for developing cost-effective advanced-technology alternatives for maintaining data continuity beyond Landsat-7 (Scheffner, 1994). To address this requirement, NASA plans to launch EO-1 as part of its New Millennium Program (Ungar, 1997). This mission will be devoted to testing new technologies for use beyond Landsat-7. Some concepts for an advanced sensor are described by Salomonson et al. (1995) and Williams et al. (1996). In essence, advanced Landsat concepts use solid-state, push-broom, multispectral linear arrays, and hyperspectral area arrays that employ grating and wedge filter technologies.

Exactly how an advanced Landsat observing capability will be implemented is still under study. One option is to fly the advanced-technology Landsat sensor on one of the NASA EOS satellites, such as the AM-2 mission. This option would reduce launch costs. Other possibilities include flying the sensor on a separate, smaller and less expensive advanced-technology spacecraft. A third possibility is for the advanced-technology capabilities and Landsat continuity requirements to be incorporated in a capability provided by a commercial entity. In any case, it is clear that the earth science and applications community needs the Landsat TM quality and type of data to be provided and continuity to be ensured so that the integrity of the databases inaugurated by Landsat-1 can be preserved. It appears clear that advanced technology can be used to meet these requirements and possibly provide highly desirable enhancements.

Table 2-4 is a chronology of Landsat and similar international satellite sys-

TABLE 2-4 Chronology of Landsat and Landsat-like Launches, 1972-2007

Year	Platform (Nationality)	Sensor
1972	Landsat-1 (United States)	MSS; RBV
1975	Landsat-2 (United States)	MSS; RBV
1978	Landsat-3 (United States)	MSS; RBV
1982	Landsat-4 (United States)	MSS; TM
1984	Landsat-5 (United States)	MSS; TM
1986	SPOT-1 (France)	HRV
1988	RESURS-01 (Russia)	MSU-SK
1988	IRS-1A (India)	LISS-1
1990	SPOT-2 (France)	HRV
1991	IRS-1B (India)	LISS-2
1992	JERS-1 (Japan)	OPS
1993	Landsat-6 (United States)	ETM
1993	SPOT-3 (France)	HRV
1993	IRS-P1 (India)	LISS-2; MEOSS
1994	IRS-P2 (India)	LISS-2; MOS
1994	RESURS-02 (Russia)	MSU-E
1995	IRS-1C (India)	LISS-3
1996	ADEOS (Japan)	AVNIR
1996	PRIRODA (Germany/Russia)	MOMS
1997	IRS-1D (India)	LISS-3
1998	CBERS (China/Brazil)	LCCD
1998	SPOT-4 (France)	HRVIR
1998	Landsat-7 (United States)	ETM+
1998	EOS AM-1 (United States/Japan)	ASTER
1998	IRS-P5 (India)	LISS-4
1999	Resource 21 (United States)	Resource 21
2000	IRS-2A (India)	LISS-4
2002	ALOS (Japan)	AVNIR-2
2002	SPOT-5A (France)	HRG
2004	IRS-2B (India)	LISS-4
2004	SPOT-5B (France)	HRG
2004	ALOS-A1 (Japan)	AVNIR-3
2007	ALOS-A2 (Japan)	AVNIR-4

NOTE: The acronyms in this table are the terms by which the platforms and sensors listed are commonly known; for the full names, see Morain and Budge (1996).

SOURCE: From Morain and Budge (1996) and Stoney et al. (1996).

tems. It lists only so-called earth resources satellites having sensors operating in the visible and near infrared spectrum, and channels roughly equivalent to those of the Landsat MSS or TM sensors. In the past 25 years there have been nearly 20 launches and 4 distinct international systems (a fifth will become operational with the launch of the China/Brazil Earth Resources Satellite). Data from these satellites are used daily by international donor agencies, government agencies at all levels, oil and mineral exploration companies, environmental consultants, value-added commercial firms, academia, and the general public. The first-order land-cover categories predictable in 1972 have expanded to include rather sophisticated higher-order applications. Continuity has been achieved in more than one sense (Morain and Budge, 1995). Use of time as a discriminant has enveloped the user community in ways that were not foreseen, and will be integral to future applications in ways that are not yet perceived. Spectral analytical procedures have evolved around the time dimension and will also be stimulated by future hyperspectral data collectors. It can truly be said that even if the Landsat program heads toward extinction, its progeny will continue to support the technology it has created.

CONCLUSIONS

The earliest visionaries, such as Jules Verne, Arthur C. Clarke, Robert Alexander, William Fisher, Archibald Park, and William Pecora, predicted great things to come as humans took their first steps into space. Of all the ventures to date, the U.S. Landsat program ranks among the most successful. Interestingly, most of the problems that have plagued this national program have not been technical, but administrative and political. Despite the difficulties related to national security issues, agency roles, delays in data delivery, funding uncertainties, and a shaky attempt to privatize a federal program, the accomplishments of the Landsat program have been extraordinary. For the 25 years from 1972 to 1997, synoptic, high-quality data have been routinely acquired, processed into an ever-improving array of digital and photographic products, and used to better measure and monitor earth resources. The Landsat series has provided new insights into geologic, agricultural, and land-use surveys, and opened new paths in the exploration of new resources. Understanding of the earth, its terrestrial ecosystems, and its land processes has been advanced remarkably through the Landsat program. Of equal importance, this program has stimulated new approaches to data analysis and academic research, and provided opportunities for the private sector to develop spacecraft, sensors, and data analysis systems and provide value-added services. It has also fostered strong international participation and a whole new generation of Landsat-like systems around the world. The political, scientific, and commercial currents over the next 25 years of earth-observing systems will be no easier to chart than were those of the first 25, but the systems that result are certain to advance human understanding and use of the planet's resources.

NOTES

1 A related, longer version of this paper appeared in *Photogrammetric Engineering and Remote Sensing* (Lauer et al., 1997).

2 The "S" in EROS was subsequently changed to stand for "System" rather than "Satellite."

3 Routine collection of MSS data was terminated in late 1992.

REFERENCES

Ashford, L.S., and J.A. Noble
    1996   Population policy: Consensus and challenges. *Consequences: The Nature and Implications of Environmental Change* 2(2):24-36.
Burrows, W.E.
    1986   *Deep Black, Space Espionage and National Security.* New York: Random House.
Colwell, R.N., editor-in-chief
    1983   *Manual of Remote Sensing: 2nd Edition.* Bethesda, Md.: American Society for Photogrammetry and Remote Sensing.
Consortium for International Earth Science Information Network
    1997   Thematic Guide: The Use of Satellite Remote Sensing. Available: http://www.ciesin.org/TG/RS/RS-home.html
Covert, K.L.
    1989   Landsat: A Brief Look at the Past, Present, and Future: PPA772, Science, Technology and Politics, Spring 1989, Syracuse University, Syracuse, N.Y.
Ehrlich, D., J.E. Estes, and A. Singh
    1994   Applications of NOAA AVHRR 1-km data for environmental monitoring. *International Journal of Remote Sensing* 15(1):145-161.
Environmental Research Institute of Michigan
    1962   *Proceedings of the First Symposium on Remote Sensing of the Environment.* Ann Arbor, Mich.: Infrared Laboratory, Institute of Science and Technology.
    1963   *Proceedings of the Second Symposium on Remote Sensing of Environment.* Ann Arbor, Mich.: Infrared Physics Laboratory, Institute of Science and Technology.
    1965   *Proceedings of the Third Symposium on Remote Sensing of Environment.* Ann Arbor, Mich.: Infrared Physics Laboratory, Institute of Science and Technology.
    1972   *Proceedings of the Eighth International Symposium on Remote Sensing of Environment.* Ann Arbor, Mich.: Center for Remote Sensing Information and Analysis, Willow Run Laboratories.
Fink, D.J.
    1980   Earth Observation—Issues and Perspectives: The Theodore von Karman Lecture. AIAA 16th Annual Meeting and Technical Display, Baltimore, Md. May 6-11. (AIAA-80-0930)
Gallo, K.P., A.L. McNab, T.R. Karl, J.F. Brown, J.J. Hood, and J.D. Tarpley
    1993   The use of NOAA AVHRR data for assessment of the urban heat island effect. *Journal of Applied Meteorology* 32(5):899-908.
Kuhn, T.S.
    1962   *The Structure of Scientific Revolutions.* Chicago, Ill.: University of Chicago Press.
Lauer, D. T.
    1990   An Evaluation of National Policies Governing the United States Civilian Satellite Land Remote Sensing Program. Ph.D. dissertation, University of California, Santa Barbara.
Lauer, D.T., S.A. Morain, and V.V. Salomonson
    1997   The Landsat Program: Its origins, evolution, and impacts. *Photogrammetric Engineering and Remote Sensing* 63(7):831-838.
Lillesand, T.M., and R.W. Kiefer
    1987   Concepts and foundations of remote sensing. Chapter 1 in *Remote Sensing and Image Interpretation*, New York: John Wiley.
Luce, C.T.
    1967   Earth Resources Observation Satellite Program (EROS)—Status and Plans. Office of the Secretary: U.S. Department of the Interior, Washington, D.C.

Mark, H.
    1984   Space Education. In *Symposium Report on Space, the Next Ten Years*, Nov. 26-28, 1984,
            United States Space Foundation, Colorado Springs, Colo.
    1988   A forward looking space policy for the USA. *Space Policy* 4(1)(Feb.).
McDonald, R.A.
    1995   CORONA: Success for space reconnaissance, a look into the cold war, and a revolution
            for intelligence. *Photogrammetric Engineering and Remote Sensing* 61(6):689-719.
McRae, M.
    1990   Fighting to save a fragile earth; man's own habitat stumbles on itself. *International Wild-
            life* 20(2)(March-April):20-21.
Morain, S.A., and A.M. Budge
    1995   Searching for continuity. *GIS World* 8(12):34.
Morain, S.A., and A.M. Budge, editors
    1996   Earth observing platforms and sensors. Vol. 2 in *Manual of Remote Sensing, 3rd Edition*,
            A. Ryerson, editor-in-chief. CD-ROM. Bethesda, Md.: American Society for Photo-
            grammetry and Remote Sensing.
Morain, S.A., and P. Thome
    1990   America's Earth Observing Industry: Perspectives on Commercial Remote Sensing. Hong
            Kong: Geocarto International Center.
Morain, S.A., and D.L. Williams
    1975   Wheat production estimates using satellite images. *Agronomy Journal* 67(3):361-364.
National Research Council
    1985   *Remote Sensing of the Earth from Space: A Program in Crisis.* Space Applications
            Board, Commission on Engineering and Technical Systems. Washington, D.C.: National
            Academy Press.
National Space Council
    1989   Landsat Issue Paper. Executive Office of the President, Washington, D.C. (April 11).
Obenschain, A.F., D.L. Williams, S.K. Dolan, and J.F. Andary
    1996   Landsat-7: Today and Tomorrow. Paper presented at the Pecora Conference, Sioux
            Falls, S.D. (August 20-22).
Office of Technology Assessment, U.S. Congress
    1993   *The Future of Remote Sensing from Space: Civilian Satellite Systems and Applications.*
            OTA-ISC-558. Washington, D.C.: Government Printing Office.
Office of the President
    1988   *Unclassified Excerpts from the New National Security Directive on National Space Policy*:
            February 8, Washington, D.C.
    1992   *National Space Policy Directive #5.* February 13, Washington, D.C.
    1994   *Presidential Decision Directive/NSTC-3 on Landsat Remote Sensing Strategy.* May 5,
            Washington, D.C.
    1996   *National Space Policy (Fact Sheet).* National Science and Technology Council, Septem-
            ber 19, Washington, D.C.
Pecora, W.T.
    1972   Remote sensing of earth resources: Users, prospects and plans. In *NASA's Long-Range
            Earth Resources Survey Program, Thirteenth Meeting.* Panel on Science and Technology,
            Committee on Science and Astronautics, U.S. House of Representatives, January 25, U.S.
            Government Printing Office, Washington, D.C.
Reeves, R.G., editor-in-chief
    1975   *Manual of Remote Sensing.* Falls Church, Va.: American Society for Photogrammetry
            and Remote Sensing.
Roller, N.E.G., and J.E. Colwell
    1986   Coarse-resolution satellite data for ecological surveys. *Bioscience* 36(7):468-475.

Sadowski, F.G., and S.J. Covington
1987   *Processing and Analysis of Commercial Satellite Image Data of the Nuclear Accident Near Chernobyl, U.S.S.R.*. U.S. Geological Survey Bulletin 1785. Washington, D.C.: U.S. Government Printing Office.

Salomonson, V.V., J.R. Irons, and D.L. Williams
1995   The Future of Landsat: Implications for Commercial Development. In *Proceedings, Conference on NASA Centers for the Commercial Development of Space*, M. El-Genk and R.P. Whitten, eds. American Institute of Physics.

Scheffner, E.J.
1994   The Landsat program: Recent history and prospects. *Photogrammetric Engineering and Remote Sensing* 60(6):735-744.

Stoney, W., V. Salomonson, and E. Schuffner
1996   *Land Satellite Information in the Next Decade: The World Under a Microscope*. Bethesda, Md.: American Society for Photogrammetry and Remote Sensing,.

Stowe, R.F.
1976   Diplomatic and legal aspects of remote sensing. *Photogrammetric Engineering and Remote Sensing*, 42(2):177-180.

Townshend, J.R.G., and C.J. Tucker
1984   Objective assessment of Advanced Very High Resolution Radiometer data for land cover mapping. *International Journal of Remote Sensing* 5(2):497-504.

Tsipis, K.
1987   Arms control treaties can be verified. *Discover* 8(4):78-91.

Tucker, C.J., H.E. Dregne, and W.W. Newcomb
1991   Expansion and contraction of the Sahara Desert from 1980 to 1990. *Science* 253:299-301.

Ungar, S.G.
1997   Technologies for future Landsat missions. *Photogrammetric Engineering and Remote Sensing* 63(7):901-905.

U.S. Department of Commerce
1980   *Planning for a Civil Operational Land Remote Sensing Satellite System: Discussion of Issues and Options*. National Oceanic and Atmospheric Administration, Satellite Task Force, June 20. Rockville, Md.: U.S. Department of Commerce.
1984   Seven bidders respond with commercialization proposals. *Landsat Users Notes* (31) (June):1-4.
1985   Award/Contract to the EOSAT Company. NA-84-DSC-00125, June 24. National Oceanic and Atmospheric Administration, U.S. Department of Commerce, Washington, D.C.

U.S. Department of Commerce and National Aeronautics and Space Administration
1987   *Space-Based Remote Sensing of the Earth: A Report to the Congress*. Washington, D.C.: U.S. Government Printing Office.

U.S. Geological Survey
1979   *Landsat Data Users Handbook*. Rev. ed. Sioux Falls, S.D.: EROS Data Center.

U.S. Geological Survey and National Oceanic and Atmospheric Administration
1984   *Landsat 4 Data Users Handbook*. Sioux Falls, S.D.: EROS Data Center.

U.S. National Aeronautics and Space Administration
1973a   Symposium on Significant Results Obtained from the Earth Resources Technology Satellite-1: National Aeronautics and Space Administration, Goddard Space Flight Center, NASA SP-327, Vol. 1 (Technical Presentations).
1973b   Third Earth Resources Technology Satellite-1 Symposium, National Aeronautics and Space Administration, Goddard Space Fight Center, NASA SP-351, Vol. 1 (Technical Presentations) 1994.

1988   Earth System Science: A Closer View: Earth System Science Committee, NASA Advisory Council, Office for Interdisciplinary Earth Studies, University Corporation for Atmospheric Research, Boulder, Colo.

Waldrop, M.M.

1982   Imaging the earth: The troubled first decade of Landsat. *Science* 215:1600-1603.

1992   *Complexity: The Emerging Science at the Edge of Order and Chaos.* New York: Simon and Schuster.

Whelan, C.R.

1985   *Guide to Military Space Programs.* Arlington, Va.: Pasha Publications, Inc.

Williams, D.L., J.R. Irons, and S.G. Ungar

1996   Landsat 7 Follow-on Mission Concepts and the New Millennium Program Earth Observer 1 Mission. Paper presented at Pecora Conference, August 20-22, 1996, Sioux Falls, S.D.

3

"Socializing the Pixel" and "Pixelizing the Social" in Land-Use and Land-Cover Change

Jacqueline Geoghegan, Lowell Pritchard, Jr.,
Yelena Ogneva-Himmelberger, Rinku Roy Chowdhury,
Steven Sanderson, and B.L. Turner II

Remote sensing—both data and image processing—and analysis through geographic information systems (GIS) are increasingly affecting the research agendas on global environmental change, as evidenced by various reports of the Intergovernmental Panel on Climate Change (IPCC) and the International Geosphere-Biosphere Programme (IGBP), as well as a number of initiatives by agencies and organizations that fund research on global change.[1] The impacts of remote sensing and GIS to date have been greatest within the environmental and policy arenas because space-based and other imagery is used primarily to determine the physical attributes of the biosphere and the earth's surface, such as forest cover or size of housing—information that is needed in spatially explicit form by various stakeholders and decision makers. The majority of the social sciences have been slow to incorporate remote sensing and GIS as basic elements of research and reluctant to respond to global-change science. The reasons are many and complex, and cannot be addressed within the scope of this chapter (see B.L. Turner, 1991, 1997a). It is sufficient to note here that the core questions of the social sciences are seen as difficult (even impossible) to address through these imaging techniques,[2] and the understanding that might be gained in those areas from spatially explicit approaches has not been fully demonstrated or appreciated.

There are now a number of opportunities to pursue some of the core social science research issues more closely through remote sensing and GIS. Examples are issues of equity/equality, gender, demography, institutions, democratization, (under)development, and decision making as they relate to resource use and environmental change. One such opportunity is represented by the core research project on Land-Use/Cover Change (LUCC) of the IGBP and the International

Human Dimensions Programme on Global Environmental Change (IHDP) (B.L. Turner et al., 1995). This project (or project framework) is designed to improve understanding of the human and biophysical forces that shape land-use/cover change through three means of assessment: (1) ground-based studies of use-cover dynamics, focused on the land manager; (2) space-based observation of the land-cover consequences; and (3) integrative models of these dynamics at various scales of analysis.

The objectives of the LUCC project include making remote sensing in general (but especially that involving satellite imagery) more relevant to the social, political, and economic problems and theories pertinent to land-use and land-cover change (B.L. Turner, 1997c, in press), which we euphemistically call "socializing the pixel" and "pixelizing the social." This objective involves methods and tools, such as GIS, that are relevant to analysis of spatial imagery and "gridded" data in general. Attempts to achieve this objective must invariably address issues of scalar dynamics—interpreting, merging, and analyzing the data and analysis across spatial, temporal, and hierarchical scales. These methods and tools and the associated scalar dynamics are the long-standing subjects of study and an extensive literature (e.g., Ehleringer and Field, 1993; Foody and Curran, 1994; Fox et al., 1995; Michener, 1994; Quattrochi and Goodchild, 1997; Rosswall et al., 1988; M. Turner, 1990; M. Turner et al., 1995; Woodcock and Strahler, 1987).[3] It is not our purpose to review this work here. It is sufficient to note that the majority of this work leads up to and has significant implications for the notion of socializing and pixelizing in land-use and land-cover change research. Most of it, however, stops short of the work that is the focus of this chapter.

To date, work on the human dimensions of global change has focused largely on indirect linkages between information embedded within spatial imagery and the core themes of the social sciences. Such work is exemplified by assessments of the proximate causes of land-use and land-cover change (e.g., slash-and-burn cultivation, clear-cutting of timber), environmental constraints/opportunities associated with human activities (soil sustainability and zones of intensive cultivation), or assessments of infrastructure in planning (e.g., green spaces, road networks) (see, for example, Behrens et al., 1994; Ehrlich et al., 1997; Fischer and Nijkamp, 1993; Martin, 1996; Massart et al., 1995; Sader, 1995; Sample, 1994; Walsh et al., 1997). This work, as important as it is, tends to address factors that mediate behavioral and social actions or outcomes, and does not focus on underlying causes and structures. Exploration of more direct linkages would involve, for example, extracting information about social standing, wealth, or human health directly from the imagery (pixels), or conducting tests of a land-use theory in which such information informs the basic tenets and operation of the underlying human model.

Various efforts are now under way to advance the notions of socializing and pixelizing, although they may not employ those terms (Entwisle et al., 1997; Frohn et al., 1996; Guyer and Lambin, 1993; Wear et al., 1996). The aim of this chapter is to clarify this genre of research on land-use and land-cover change, illustrate the work through a few select examples, identify some of the scalar issues confronting research of this kind, and present some major lessons this research has begun to reveal.

SOCIALIZING THE PIXEL

As suggested earlier, to socialize the pixel is to take remote sensing imagery beyond its use in the applied sciences and toward its application in addressing the concerns of the social sciences. Two avenues of research of this kind offer the potential to shed light on land-use and land-cover change: mining the pixel and modeling from the pixel.

Mining the pixel involves seeking social meaning in imagery—information and indicators relevant to such concerns as economic well-being or "criticality" (Kasperson et al., 1995), perhaps signaling the underlying processes that give rise to land-use and land-cover change. This meaning is often hidden deep within the analysis of the imagery (Moran et al., 1994), and this very depth may impede such investigation. Two examples illustrate this concept.

Work in landscape ecology indicates that landscape patterns indicative of gross operating processes, some social in origin, can be identified with the use of spatially explicit indicators and fractal analysis (O'Neill et al., 1988). Might similar patterns be found through the use of remotely sensed data? Mertens and Lambin (1997) use Landsat Thematic Mapper (TM) imagery and GIS analysis to identify at least six spatial patterns of land-use and land-cover change in eastern Cameroon that are indicative of market, subsistence, policy, and urbanization processes, although the authors do not follow the pattern-process linkage fully. A rudimentary attempt to do so using TM imagery for Nepal proved difficult (Millette et al., 1995). These landscapes are a composite of rain-fed and irrigated agriculture, village forests, household trees, small grazing plots, and landslide features. Changes in the mosaics or composite patterns of these land covers appeared to signal general trajectories of the socioeconomic health and environmental sustainability of the production system at the village or village cluster level, although the sample was not sufficient to determine the statistical significance of this inference. Nevertheless, work by Moran and colleagues (1994; Mausel et al., 1993) demonstrates the details of human action that can be found in pixel analysis and, coupled with the inferences from the Nepal work, suggests that further explorations of this kind are warranted.

As a second example, use of time series and principal component analyses—long used in other research—with remote sensing and GIS is gaining momentum, and this has implications for social science research (e.g., Anyamba and Eastman, 1966). For instance, a time series analysis of Advanced Very High Resolution Radiometer (AVHRR) imagery of a forest reserve in Malawi provides strong Normalized Difference Vegetation Index (NDVI)-based evidence of decreasing forest biomass, apparently because residents adjacent to the reserve trim the trees for fuel (Eastman and Toledano, 1996). This activity, of course, has significant implications for energy consumption, human health, policy enforcement, and economic well-being. The evidence, however, is found in the seventh residual (component) of the analysis, which is usually ignored in such analyses. Nonetheless, in the study of land-use and land-cover change, the reality of the component is an empirical question because the identified human imprint can actually be observed (or not).

These two examples are surely not exhaustive of avenues of research that offer the potential to make remote sensing more relevant to social science interests. They illustrate, however, that a business-as-usual approach in the remote sensing community may not be sufficient for serving these interests. More attention must be paid to the less obvious signals in the imagery, be they complex arrays in the patterns of land cover or changes found deep within the analysis of land-use and land-cover change.

In addition to the search for social meaning in the pixel, various approaches can be used to model from the pixel toward the interests of the social sciences, although they have been minimally explored. These approaches are largely empirical and atheoretical in nature, but can be used to model land-use and land-cover change directly from remotely sensed imagery. Markov chain modeling, for example, offers a means of addressing land-use and land-cover change when the data are not spatially explicit enough or are of spatially coarse resolution. More important, however, it allows an inductive exploration of land-use and land-cover change that may provide clues about the underlying dynamics involved.

Markovian approaches assume that the immediate past is the best predictor of the near future, under the condition of stationarity, and uses transition probabilities of past states (e.g., land uses or covers or their signals in the imagery) to estimate future states. Such approaches have been used successfully for estimating change in phenomena involving processes that conform to the stationarity principle, where previous land uses are a proxy for stationary human behavior (Usher, 1981). These approaches would seem appropriate for cases involving low levels of chronic change, as in the case of subsistence-driven cultivation in forest regions that is associated with natural population growth or decline.

Unfortunately, most cases of land-use and land-cover change do not conform to the stationarity principle. Rather, changes are the result of multiple actors and structures combining in complex, synergistic ways. Moreover, critical exogenous forces, especially international and national policy decisions, may have profound effects on land-use and land-cover change. These forces can be seen as shocks to the existing land management system that fundamentally alter the pathways and trajectories of change, thus rendering the estimated probabilities from Markovian and other such analyses invalid. To address some of these problems, "raw" Markov applications drawing on remotely sensed data, such as the pixel, can be socialized by accounting for various land-use and land-cover factors that change the estimated probabilities in question: soil quality or slope, management rules, resource institutions, and so on. Where this understanding can be registered to the pixel, that is, where these attributes can be recorded along with the pixel's geographic and spectral characteristics, the estimated probabilities of change from observations of the past can be altered. In this process, of course, a point is reached at which the atheoretical empirical approach is affected by the choice of the variables incorporated in the analysis, complete with their theoretical implications.

An exploration of this kind is under way as part of a Southern Yucatán Peninsular Region (SYPR) project that will produce, compare, and merge Markov modeling approaches based on remote sensing with field- and statistics-based models (see below) of the various land managers that are producing the signals registered in the imagery.[4] SYPR focuses on the southern portions of the Mexican states of Quintana Roo and Campeche, extending along Route 186 north of the border of Petén, Guatemala. The dominant semideciduous tropical forests of the region came under assault after the highway was built, becoming the pathway for various new land users: first were slash-and-burn farmers on communally designated lands, followed by private ranchers, rice projects sponsored by nongovernmental organizations (NGOs), and, more recently, various smallholder market operations and the Calakmul biosphere reserve. Virtually the entire period of major land-use and land-cover change is captured by the Landsat Multispectral Scanner (MSS) and TM.

A pretest exploring the socialization of a Markov approach has been undertaken with MSS imagery alone as a precursor to the larger SYPR effort, and is now is in its final stages. To model from the pixel as argued here, the Markovian transition probabilities of land-use and land-cover change must be made spatially explicit, and expanded through the insertion of biophysical and socioeconomic factors into the probability analysis. For the pretest, three Landsat MSS images spanning the period 1975 to 1990 were used, focusing on the southern Yucatán peninsular region and more or less centered on Lago Silvituc, Campeche. Six land-cover classes were derived in three land-cover maps (1975, 1986, and 1990) produced by supervised classification of MSS imagery: forest, scrub vegetation (bush and early secondary growth), natural savanna, land in crops, bare soil/roads, and water. These land-cover maps were overlaid in a GIS to create transition maps for the two periods 1975-1986 and 1986-1990. By calculating the transition probability of each cell in the land-cover maps as a function of existing land covers in the neighborhood of that cell, a spatial component was added to the transition probabilities. Multinomial logistic regressions were used to link the spatially explicit actual land-use transitions to biophysical, distance, and socioeconomic variables. A suite of models involving different combinations of explanatory variables, some of which begin to socialize each cell (pixel), were estimated for the transition types from the map for the first period (1975-1986), and the estimated coefficients from each model were used to predict land-use changes in the following time period. These predicted probabilities of transition were then compared with actual transitions for the second period (1986-1990).

Because the basic transition during the study period was deforestation for cropland (with a minor amount for pasture), we focus here on two transitions: persistence in forest cover and conversion of forest cover to cropland. Using the spatially explicit Markov approach, the suite of models correctly predicts 94.5 to 96.5 percent of the observed persistence in forest cover for the transition period 1986-1990. This finding is not surprising given that the overwhelming land cover was forest sufficiently distant from human activity to be protected from conversion. The same suite of models was less accurate in predicting transitions from forest to cropland. In this case, the raw Markov model predicted only 16.2 percent of the observed transitions (1986-1990) correctly, but up to 20.0 percent predictive accuracy was achieved by including distance variables (to roads, villages, and markets).

These results may not seem so promising, but the conditions of the trial must be understood. First, no assessment of the classification accuracy of the identified land covers was undertaken. Second, only three social variables were used because of inadequate availability and consistency, and the surrogates used for village affluence (number of cattle and trucks) may have been inappropriate. However, most of the transition probabilities were generated for a period during which large-scale, state-sponsored agricultural projects were undertaken along the new highway, biasing transition probabilities toward high rates of forest conversion to cropland. The predicted period (1986-1990), however, witnessed a collapse of this dynamic. In short, the assumed principle of stationarity was violated.

Given this result, one might conclude that Markovian approaches of this kind may prove problematic and not useful. Yet examination of the ability of the expanded Markov models to explain the transitions found during period one (1975-1986) indicates that improvement is gained by socializing the pixel. Using the same suite of models (increasing in complexity with the addition of biophysical, distance, and socioeconomic variables), with a measure of the fit of the model increasing from 7.0 to 34.2 percent (pseudo R^2). This result suggests that further exploration of the socialized Markovian approach may be useful, especially with superior data and techniques accounting for time lags and shocks to land-conversion dynamics.

PIXELIZING THE SOCIAL

A paucity of spatially explicit data has constrained spatial modeling of human behavior and social structures, especially beyond the field of geography, and fostered modeling approaches that abstract from the essential spatial nature of the problem. As a result, either aggregate relationships are specified, or the spatial components in a model are reduced to unidimensional variables, such as the distance between economic activities in a location model, the wage differential in a migration model, or the cost of access in a transportation model. The increased availability of spatially explicit data, both remotely sensed and other data, and GIS has begun to change this situation, especially with regard to broadly interpreted land-use and land-cover change. Advances are being made in linking on-the-ground human actions and consequences to imagery (pixels) through models, or modeling to the pixel. Such efforts require, for our interests here, that each pixel (or gridded datum) be modeled to have an empirically estimated probability of change from one land use or land cover to another. In contrast to modeling from the pixel as discussed earlier, such estimates are derived by linking theoretical models directly to the imagery (Lambin, 1994), as in modeling the determinants of the decisions of individual land managers on the basis of utility maximization, satisficing, or other theories of human behavior.

Related to these approaches are questions of empirical estimation and empirical tests of hypotheses of human behavior or social structures using remotely sensed data. There appears to be a general belief among some social scientists that the uses of spatial data are restricted to enhancing the measurement and definition of explanatory variables. If this were true, the value of geographical (spatially explicit) data for the social sciences would certainly be an empirical issue and would be sensitive to the characteristics of the particular application. Better data would indeed yield higher payoffs. There are, however, additional potential gains to be realized from using spatial data. Analyzing a problem that is essentially location based without geographically coded data is analogous to analyzing a time-series problem without knowing the chronological order of the observations.[5] The further development of statistical techniques for estimating spatially explicit models using remotely sensed data is essential, as articulated for spatial econometrics by Bockstael (1996). Taking account of the spatial nature of the problem will improve estimation and shed light on interactions and interdependencies in the system that may be interesting in their own right.

Additionally, what we euphemistically refer to as mining and modeling the pixel can be brought together in land-use and land-cover change studies. Creative explanatory variables can be constructed from remotely sensed data through the use of GIS, as with landscape patterns or land-use mosaics. If these patterns and mosaics (e.g., fragmentation, urbanization) significantly change land-management options, one could hypothesize a relationship between changing patterns in an area over time and explore the effect of that relationship on an individual land owner's current management schemes (e.g., Frohn et al., 1996). For example, using an index of the pattern as an additional explanatory variable in a model, the relationship between past and current patterns and current land-use decisions can be estimated (Geoghegan et al., 1997). However, to include such variables in an empirical specification, a model must start with a theoretical understanding of the human behavior of valuing different types of land uses and the distribution of those land uses across the landscape (Geoghegan and Bockstael, 1997).

Human-induced land-use and land-cover change is currently being modeled for the data-rich Patuxent Watershed of the Chesapeake Bay,[6] revealing the spatial configuration and dynamic evolution of a landscape by capturing ecological functions, human behavior, and their interaction. The effort links remotely sensed data on land use and land cover with a variety of spatially explicit socioeconomic and physical data, as well as with separate ecological and economic models that are constructed so that the outputs of each can be easily used as inputs to the other. Each model employs a landscape perspective that captures the spatial and temporal distributions of the services and functions of the natural system and human-related phenomena, such as surrounding land-use patterns and population distributions. Configuration and reconfiguration of the landscape follows from the intertwining of these phenomena, and the Patuxent work offers the potential for a richer model of land use and its change by accounting for spatial heterogeneity and linking land-use conversion to features of the landscape. The aim is to predict the probability that a given pixel of a given description and in a given location will remain in its current use or be converted to an alternative use. While the conversion process is affected by inertia and other disequilibrium considerations and constrained by zoning and other land-use controls, the changes in land-use probabilities are functions of the value of each parcel in alternative uses. Consequently, the analysis must be able to explain what factors affect land values in alternative uses (Bockstael, 1996).

The land within the 7,000 km^2 of seven counties of the Patuxent Watershed located in Maryland ranges from suburbs of Washington, D.C. to rural and agricultural areas of southern Maryland. The conversion of agricultural and forested land (open use) to residential uses constitutes 78 percent of the total land-use change in these seven counties during the past 10 years. As a consequence, the economic modeling effort focuses on prediction of the conversion of open land use to residential use through a four-part process: (1) analysis of residential value as function of a variety of spatially related economic and ecological variables that are hypothesized to affect residential land values,[7] estimated on actual transactions of residential parcels; (2) use of the estimated coefficients of the explanatory variables from step 1 to predict values for open land of a given description were it to be converted to residential use; (3) use of these predictions with other explanatory variables, such as zoning, soil type, and costs of conversion, to estimate the spatial distribution of the relative probability that any such land will be developed; and (4) linking of these relative probabilities with a macroeconomic model of the state of the local economy to predict annual housing starts and thereby how much land (how many of the pixels) will change in a given year.

Many of the themes noted earlier are included in this ongoing modeling endeavor. The model is a utility-theoretic econometric model of human behavior affecting land-use decisions, not driven by GIS determinism. Using spatial data leads to interesting complications, such as spatial autocorrelation, temporal dynamics, and spatial structural change. Therefore, applying standard econometric techniques to either aggregate or disaggregate spatial data generates nonspherical disturbances, misspecification, and measurement error. New estimation techniques in spatial econometrics have been developed to take some of these issues into account in the Patuxent modeling effort (e.g., Bell and Bockstael, 1997). Whether the information gained by using spatial econometric techniques vastly improves the estimation is still an empirical issue. The initial spatial econometric modeling work with the Patuxent model demonstrates, however, the potential improvements in explaining and predicting land values (Geoghegan and Bockstael, 1995). Further improvements and refinements of both theoretical and applied econometric modeling techniques for use with the Patuxent model are presently under way.

Another theme introduced above is not just linking the pixel to people, but trying to use the remotely sensed data more creatively. For example, to better capture the spatial externalities that often characterize land use and also influence land values, indices based on the diversity and fragmentation of the surrounding landscape around each pixel have been included in the Patuxent land-value model to further explain residential land values. The intuition behind including these variables is that increasing land-use and landscape diversity may adversely affect aesthetics, but may also have convenience value signifying the proximity of important work, shopping, recreation, and institutional destinations. Which effect dominates is an empirical question. Fragmentation might be considered more obviously undesirable. Holding diversity constant, increasing fragmentation signals a hodgepodge of land uses. A high fragmentation index is synonymous with a checkered landscape, and implies the potential for large negative locational externalities. Confusion over the sign of expected effects may be very much tied to the issue of scale (see below). Preliminary estimation demonstrates that these additional GIS-created variables, measured at different scales, can add explanatory power to the Patuxent model of housing values (Geoghegan et al., 1997). Not only do these variables add explanatory power, but, depending on the scale at which they are measured, the spatial index variables of land use can be either amenities (adding to value) or disamenities (reducing value). For example, the estimated coefficient on a small-scale measure of diversity implied that individuals valued a homogenous pattern of land use in their immediate neighborhood, but the estimated coefficient of a larger-scale diversity measure implied that a higher degree of heterogeneous land uses was valued at this higher scale. The nature and pattern of the land uses surrounding a parcel have an influence on the price, implying that people care very much about the patterns of landscape around them, and supporting the belief that severe externalities exist in land use and land-use patterns.

An illustrative application of this model (Bockstael and Bell, 1997) involved steps 1-3 above (see Figure 3-1). This map shows the outcome of a model

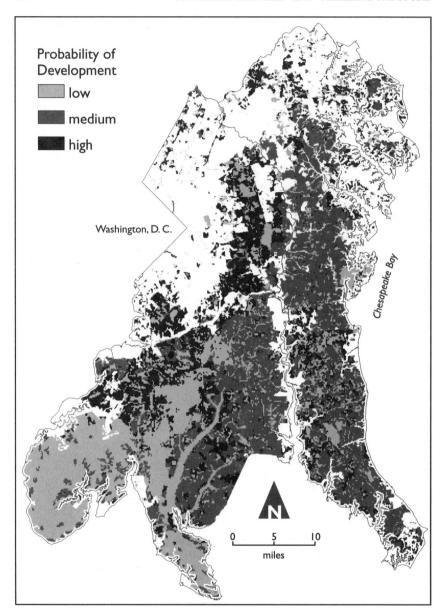

FIGURE 3-1   Predicted probability of development: Anne Arundel, Calvert, Charles, and Prince George's counties, Maryland.

focusing on the four southern counties of the Patuxent Watershed.[8] Areas more darkly shaded have a higher probability of development; all nonshaded areas are land parcels that are either currently developed or precluded from development. Some of the areas of higher predicted probability of development are closer to Washington, D.C., as would be predicted by a traditional model of residential choice, whereby commuting distance is an important cost component in individuals' choices of residential location. Other areas with high probability of development are waterfront properties, as would be predicted by a simple spatial amenity model. However, it is interesting to note from this map that even after controlling for these two effects, there are still many areas dispersed throughout the region that have a high probability of development.

Given the spatially disaggregate data and GIS capabilities, hypotheses were developed to test how the distribution of land uses around a location can affect human behavior—specifically, how individuals value this distribution of land uses and how these values can affect probabilities of land-use change. The estimated econometric results on which Figure 3-1 is based suggest that individuals do value different types of land-use patterns and seek to reside in locations that have a specific distribution of local land uses, which show up as the darker areas on Figure 3-1 in regions that are not waterfront property or relatively close to Washington, D.C. Through this modeling exercise and earlier work on including spatial land-use indices in econometric models (Geoghegan et al., 1997), it was found that including dissaggregate location-specific data to control for different amenities and disamenities in land use greatly enhanced the models' explained variance. Adjusting for spatial statistical issues also enhanced prediction (Bell, 1997).

SCALAR DYNAMICS AND PATH DEPENDENCE

In the Patuxent, SYPR, and other analyses, land covers are modeled as a function of biophysical and socioeconomic variables and their interactions. The critical variables change in incidence and importance, however, through time and across scales of analysis (Sanderson and Pritchard, 1993). A primary challenge in the remote sensing and GIS initiative to model these variables is to escape the tendency toward a GIS-driven determinism, which tries to account for land-cover heterogeneity through elaborate map algebra involving multiple layers of landscape features. The landscape is commonly taken to be in some kind of dynamic equilibrium: driving forces, human or not, may change, creating a kind of disturbance, but endogenous processes restore the equilibrium. Even within this framework, land use and land cover are often not a simple function of these endogenous processes; there can be time lags and spatial-diffusion processes, and the processes themselves can be buffered, amplified, inverted, or otherwise transformed before the resulting change in the landscape can be seen.

As an example, consider the case of the proposition that international agricultural prices determine a significant share of agricultural land use. To what extent (and through what mechanisms) do these prices pass through to the micro level? Is it true that all hierarchical systems conduct price signals with identical "resistance"? If not, what are the implications for using international prices as a driving force at the unit of production (land-use manager) scale? Most important, what makes different land-use systems more or less permeable to such macroscale signals? How do these land-use system vulnerabilities vary through time?

As another example, many researchers have found strong links between the external sector (international commodity prices and exchange rate dynamics, or El Niño/Southern Oscillation [ENSO] phenomena) and changes in land use or land cover, such as forest biomass or cropping schedules. These results are difficult if not impossible to generalize across regions and nations, however, and simple correlations tend to fall apart (for a review see B.L. Turner et al., 1995). Similarly, some research postulates a straightforward link between population levels (or rates of change) and deforested area (or deforestation rates), but such relationships typically explain no more than 50 percent of the variance in forest cover across diverse regions (Mather, 1996) and commonly disappear in place-specific analysis (Kasperson et al., 1995). When these supposed macro-mechanisms are not understood and set in context, even statistically significant correlations may be spurious. This simple but sometimes overlooked observation is one of the major scaling issues that confronts modeling efforts.

Even when these lags and transformations are taken into account, land-use and land-cover systems do not always respond in predictable ways to predicted driving forces because land cover is a function not only of socioeconomic and biophysical variables, but also of itself. A mismatch between driving forces and the state of land cover is not necessarily an indication that the scale of analysis is wrong. Endogenous, contagious processes (e.g., fire, disease outbreak, techno-logical change and diffusion, or frontier clearing) may well explain the break-down in predictive capacity between scale levels. It may be important to consider the extent to which a linked land-use and land-cover system exhibits its own dynamic (even apart from driving forces). If the system is path dependent, its current state and trajectory of change depend on its history (as in Markov ap-proaches noted above), not on current values of driving forces alone.

A path-dependent system may exhibit several properties that must be consid-ered in land-use and land-cover change assessments (Arthur, 1989): varying predictability (unpredictability is followed by high predictability as the system is "locked in"); nonergodicity (historical events are not averaged away, and small perturbations may significantly influence long-run development); progressive inflexibility (the system is ultimately insensitive to perturbation); and potential inefficiency (the outcome is not optimal for society).

Path dependency may arise from several sources, but two primary sets of dynamics are discussed in the literature. The first, self-reinforcement, is a pro-cess of increasing returns to scale (David, 1985) or network externalities, a form of increasing returns to agglomeration (Arthur, 1995; Krugman, 1994). The geographic concentration of industries, the ascendancy of particular agricultural technologies, and the dynamics of commodity booms and busts are all explicable in these terms. Historical accident may explain as much as driving forces do. The other set of processes leading to path dependency is investment rigidities— sunk costs, infrastructure development, landesque capital such as field drains and terraces, institutional evolution—that constrain and shape future development possibilities. The two are not mutually exclusive: self-reinforcing processes build their own infrastructure, leading to irreversibility and inflexibility (at least in the short run). Path dependencies and disequilibria in land-use change are also currently being explored as part of the larger Patuxent (Irwin, 1997) and SYPR modeling projects described above.

Recognition that historical accident is critical to land-use and land-cover analysis does not preclude modeling, projecting, or other such scientific efforts. As important as path dependence may be, it does not subsume all land dynamics. The task is to identify when and where it operates and thus the spatial, temporal, and hierarchical scales in which general dynamics operate.

Why not just choose a different scale for modeling—one that spans these historically contingent processes, or one that encompasses and subsumes location-specific heterogeneity? This strategy has, in fact, been recommended for modeling based on traditional hierarchy theory. Fine-scale unpredictable processes are seen as being filtered out at higher levels of organization, so that they appear as averages or statistical distributions, with details smoothed out or aggregated. On the other hand, broad-scale processes are so slowly changing and infrequent that they appear as constraints in a model. Such "vertical decoupling" of time dynamics allows models to be built at a small range of scales, without having to be concerned about cross-scale interactions. Scaling up and scaling down thus require nesting models together and specifying the weak linkages among them (Pattee, 1973).

However, land-use and land-cover change is not a single hierarchy of processes along a continuous space-time graph, so capturing the necessary effects and dynamics is not just a question of finding the right scale of analysis. Parallel hierarchies of geological/edaphic conditions, human land-use processes, vegetative processes, and atmospheric dynamics exist (Gallopín, 1991). Scaling up and down within each hierarchy is one matter, but linking across hierarchies may occur for processes with similar spatial scales and very different time dynamics, which makes the neat predictions from the vertically decoupled world collapse.

Even within a hierarchy, levels are not static, and mechanisms for transmitting cross-scale signals are not stable. When fine-scale processes are linked together, they may give rise to sudden changes, to radical flips between alternate stable states in systems. For example, long periods of stability in relative prices may synchronize previously heterogeneous agricultural systems across a landscape. During this period, there may be a relatively strong relationship that explains incremental change in land-use systems as a function of small changes in relative commodity prices. But the very process of synchronization may create vulnerability to a previously unimportant variable—lightning strikes, an agricultural pest, or rainfall variability—which may have existed previously, but now has a much broader-scale effect than in the heterogeneous system. Synchronization leads to a progressive loss of resilience, defined by the size of the perturbation a system can tolerate and still recover (Holling, 1986).[9]

Such surprises occur over multiple scale ranges, so predicting when the models will break down is as necessary as predicting change at particular scales. By exploring the interactions between slow variables (e.g., the gradual synchronization of agricultural systems in a market) and fast variables (e.g., pest outbreaks, fires), understanding can be directed toward the limits to predictability and the generation of surprise (Sanderson and Holling, 1996).

CONCLUSION

The evolution of global environmental change to global sustainability (B.L. Turner, 1997a) has enlarged the human dimensions of the research agenda, increasingly necessitating cooperation and collaboration among the natural, social, and remote sensing/GIS sciences. The LUCC project and initiatives within the project that involve socializing the pixel and pixelizing the social offer the potential to achieve such integration. They do so because they do not regard the social sciences as an appendage to the natural and remote sensing/GIS sciences, but hold the promise of providing information and understanding that speak to the core issue of social science understanding.

While largely in its infancy, work of this kind provides some rudimentary lessons that warrant special attention for the various collaborative research initiatives under way:

• Indicators of social or human-environment conditions in remotely sensed data, especially satellite imagery, are likely to be found in complex and composite patterns, requiring analytical techniques and tools to register. Since these patterns are generated by the unfolding of many processes in place, as in the Nepalese case, they are likely to be applicable only at the regional level.

• Depending on the signal in question, as the Malawi case suggests, the role of seasons and climatic flux may mask the human imprint. Analytical means must be employed to filter through the layers of information in the signal to find the human imprint of spatial processes.

• Probability approaches from the pixel per se can be made spatially explicit to meet the needs of the land-use and land-cover change research community.

• Markov chain and other such probability approaches have not been sufficiently explored in the sense of socializing the pixel, and the conditions under which they may provide robust modeling outcomes (or not) remain unclear.

• As social scientists use spatial data, their theoretical models of human behavior need to be made more spatially explicit to aid understanding of how humans use and value the landscape at different scales.

• As empirical specifications of models are developed using remote sensing and GIS, development of the necessary statistical techniques for estimation is critical.

• Ultimately, various issues of scalar dynamics must be addressed and resolved if improvement in understanding and modeling of land-use and land-cover change is to be achieved. These issues include buffering, amplification, and inversion of transforming processes; the path dependence and historicity of process-system relationships; and coupling of synchronous processes with parallel hierarchical structures.

ACKNOWLEDGMENTS

This paper explicates various themes under development by the International Geosphere-Biosphere Programme (IGBP)-International Human Dimensions Programme on Global Environmental Change (IHDP) core project on Land-Use/Cover Change (LUCC), although it is not a formal LUCC document. Sanderson and Turner serve on the Science Steering Committee (SSC) LUCC, and Sanderson and Pritchard are, respectively, the chair and science officer of the LUCC Focus 1 Research Activity. The authors thank the SSC LUCC and various colleagues at the University of Florida and University of Maryland and in the George Perkins Marsh Institute, Clark University, for their insights. Parts of this paper were supported by the Carnegie Mellon University Center for Integrated Studies (NSF-SBR 95-21914), the U.S. Man and Biosphere Program (Tropical Ecosystems Directorate, #TEDFY94-003), the U.S. Environmental Protection Agency (#CR8219525010 and #R825309-010), and the Maryland Agricultural Experiment Station (AREC-96-62).

NOTES

1 See, for example, IGBP Report 35/IHDP Report 7 (Turner et al., 1995) or the 1995 IPCC report (Houghton et al., 1995).

2 Variously articulated, these questions can concern humanity's relationship with the mystical and religious, with itself, and with nature (B.L. Turner, 1997b). With the exception of the last question, immediate links to remote sensing are not necessarily apparent. More important, however, understanding in the social sciences is embedded in human behavior and social structures, the essence of which is not readily linked to remote sensing and, until recently, not commonly conceived in terms of spatial relations (see National Research Council, 1997).

3 It is interesting to note that many advances made in spatial geography during the 1960s and 1970s, which gave way to interdisciplinary spatial statistics and analysis more generally, are likewise applicable to the stated objective (National Research Council, 1997). Unfortunately, many of the research communities now engaged in examining the human dimensions of global change are largely

unaware of lessons learned from these efforts, and the community of these spatial researchers has not synthesized these lessons in ways that make them tractable for other communities.

4 SYPR is funded by the National Aeronautics and Space Administration's Land-Cover and Land-Use Change initiative (NASA-NAG 56046) and involves a collaboration of the George Perkins Marsh Institute (Clark University), Harvard Forest (Harvard University), and El Colegio de la Frontera Sur—Unidad Chetumal with the assistance of Carnegie Mellon University Center for Integrated Studies (NSF-SBR95-21914).

5 While geographers have long championed the significance of place and spatial relations for understanding (National Research Council, 1997), this significance is increasingly embedded within critical social postmodern approaches to understanding, and often takes the form of "context" or "contextualization."

6 This project is funded by the U.S. Environmental Protection Agency (Cooperative Agreement #CR82 19525010), the Maryland Agricultural Experiment Station (AREG-96-62), and the EPA/NSF Decision Making and Valuation for Environmental Policy Research Initiative (EPA Grant #R825309-010), and involves a collaboration between Clark University and the University of Maryland.

7 These values are predicated on such attributes as location, distances to features in the landscape, view, and surrounding landscape amenities and neighboring land uses, where the land use is residential or another developed use. In the case of residential land values, individuals are modeled to trade off reduced commuting distance to major employment centers for lot size, as well as neighborhood and environmental amenities.

8 Anne Arundel, Prince George's, Calvert, and Charles counties.

9 Such flips or collapses in land-use and land-cover systems are the subject of research for the emerging Resilience Network of the Beijer Institute for Ecological Economics, Stockholm.

REFERENCES

Anyamba, A., and J. R. Eastman
    1996   Interannual variability of NDVI over Africa and its relation to El Niño/southern oscillation. *International Journal of Remote Sensing* 17:2533-2548.
Arthur, W.B.
    1989   Competing technologies, increasing returns, and lock-in by historical small events. *The Economic Journal* 99:116-131.
    1995   Urban systems and evolution. Pp. 1-14 in *Cooperation and Conflict in General Evolutionary Processes*, J.L. Casti and A. Karlqvist, eds. New York: John Wiley.
Behrens, C.A., M. G. Baksh, and M. Mothes
    1994   A regional analysis of Bari land use intensification and its impact on landscape heterogeneity. *Human Ecology* 22: 279-316.
Bell, K.P.
    1997   A Spatial Analysis of the Transportation-Land Use Linkage. Ph.D. dissertation. Department of Economics, University of Maryland.
Bell, K.P., and N. E. Bockstael
    1997   Applying the Generalized Method of Moments Approach to Spatial Problems Involving Micro-Level Data. Department of Agricultural and Resource Economics Working Paper #97-03, University of Maryland.
Bockstael, N. and K. Bell
    1997   Land use patterns and water quality: The effect of differential land management controls. In *International Water and Resource Economics Consortium, Conflict and Cooperation on Trans-Boundary Water Resources*, Richard Just and Sinaia Netanyahu, eds. Kluwer Publishing.

Bockstael, N.E.
  1996   Modeling economics and ecology: The importance of a spatial perspective. *American Journal of Agricultural Economics* 78:1168-1180.
David, P.A.
  1985   Clio and the economics of QWERTY. *American Economic Review* 75:332-337.
Eastman, J.R., and J. Toledano
  1996   Forest monitoring in Malawi using long time series vegetation index data. *Earth Observation Magazine.* 5(9):28-31.
Ehleringer, J.R., and C.B. Field, eds.
  1993   *Scaling Physiological Processes: Leaf to Globe.* San Diego, Calif.: Academic Press.
Ehrlich, D., E.F. Lambin, and J-P. Malingreau
  1997   Biomass burning and broad-scale land-cover changes in western Africa. *Remote Sensing of Environment* 61:201-209.
Entwisle, B., R. Rindfuss, S. Walsh, T. Evans, and S. Curran
  1997   Geographic information systems, spatial network analysis and contraceptive choice. *Demography* 34:171-187.
Fischer, M.M. and P. Nijkamp, eds.
  1993   *Geographic Information Systems, Spatial Modelling, and Policy Evaluation.* Berlin, Germany: Springer-Verlag.
Foody, G.M., and P.J. Curran, eds.
  1994   *Environmental Remote Sensing from Regional to Global Scales.* New York: John Wiley & Sons.
Fox, J., J. Jrummel, S. Yarnasarn, M. Ekasingh, and N. Podger
  1995   Land use and landscape dynamics in northern Thailand: Assessing change in three upland watersheds. *Ambio* 24:328-334.
Frohn, R.C., K.C. McGwire, V.H. Dale, and J. E. Estes
  1996   Using satellite remote sensing analysis to evaluate a socio-economic and ecological model of deforestation in Rondonia, Brazil. *International Journal of Remote Sensing* 17:3233-3255.
Gallopín, G.C.
  1991   Human dimensions of global change: Linking the global and local processes. *International Social Sciences Journal* 130:707-718.
Geoghegan, J. and N. Bockstael
  1995   Economic Analysis of Spatially Disaggregated Data: Explaining Land Values in a Regional Landscape. Paper presented to the Association of Environmental and Resource Economists at the Allied Social Sciences Association meeting, Washington, D.C., January 1995.
  1997   The Value of Open Spaces for Residential Land Prices and Land Use Change. Working paper, Department of Economics, Clark University.
Geoghegan, J., L. Wainger, and N. Bockstael
  1997   Spatial landscape indices in a hedonic framework: an ecological economics analysis using GIS. *Ecological Economics* 23:251-264.
Guyer, J., and E. Lambin
  1993   Land use in the urban hinterland: Ethnology and remote sensing in the study of African intensification. *American Anthropologist* 95:839-859.
Holling, C.S.
  1986   The resilience of terrestrial ecosystems: local surprise and global change. Pp. 292-317 in *Sustainable Development of the Biosphere,* W.C. Clark and R.E. Munn, eds. Cambridge, England: Cambridge University Press.

Houghton, J.T., L.G. Meira Filho, B.A. Callender, N. Harris, A. Kattenberg, and K. Maskell, eds.
1995   *Climate Change 1995: The Science of Climate Change.* Cambridge, England: Cambridge University Press.

Irwin, E.
1997   Local Interactions and Aggregate Effects: The Role of Spatial Externalities in the Evolution of Land Use Pattern. Dissertation proposal, Department of Agricultural and Resource Economics, University of Maryland.

Kasperson, J.X., R.E. Kasperson, and B.L. Turner II, eds.
1995   *Regions at Risk: Comparisons of Threatened Environments.* Tokyo: United Nations University.

Krugman, P.R.
1994   Complex landscapes in economic geography. *American Economic Review* 84:412-416.

Lambin, E.
1994   Modelling deforestation processes: A review. *Trees Series* B/11. Luxembourg: European Commission DGXIII.

Martin, D.
1996   *Geographic Information Systems: Socioeconomic Applications,* 2nd ed. New York: Routledge.

Massart, M., M. Petillon, and E. Wolff
1995   The impact of an agricultural development project on a tropical forest environment: The case of Shaba (Zaire). *Photogrammetric Engineering and Remote Sensing* 61:1153-1158.

Mather, A.
1996   The Human Drivers of Land-Cover Change: The Case of Forests. Paper presented at the Open IGBP/BACH-LUCC Joint Inter-Core Projects Symposium on Interactions between the Hydrological Cycle and Land Use/Cover. Kyoto, Japan, Nov. 4-7.

Mausel, P., Y. Wu, Y. Li, E. Moran, and E. Brondizio
1993   Sectral identification and successful stages following deforestation in the Amazon. *Geocarta International* 8:1-11.

Mertens, B., and E. F. Lambin
1997   Spatial modelling of deforestation in southern Cameroon. *Applied Geography* 17(2):143-162.

Michener, W.K., ed.
1994   *Environmental Information Management and Analysis: Ecosystem to Global Scales.* London, England: Taylor & Francis.

Millette, T.L., A.R. Tuladhar, R.E. Kasperson, and B.L. Turner II
1995   The use and limits of remote sensing for analyzing environmental and social change in the Himalayan middle mountains of Nepal. *Global Environmental Change* 5:367-380.

Moran, E., E. Brondizio, P. Mausell, and Y. Wu
1994   Integrating Amazonia, land use and satellite data. *BioScience* 44:329-338.

National Research Council
1997   *Rediscovering Geography: New Relevance for Science and Society.* Rediscovering Geography Committee. Washington, D.C.: National Academy Press.

O'Neill, R.V., J.R. Krummel, R.H. Gardner, G. Sugihara, B. Jackson, D.L. DeAngelis, B.T. Milne, M.G. Turner, B. Zygmunt, S.W. Christensen, V.H. Dale, and R.L. Graham
1988   Indices of landscape pattern: *Landscape Ecology* 1:153-162.

Pattee, H.H.
1973   The physical basis and origin of hierarchical control. Pp. 73-108 in *Hierarchy Theory: The Challenge of Complex Systems,* H.H. Pattee, ed. New York: George Braziller.

Quattrochi, D.A. and M.F. Goodchild, eds.
1997   *Scale in Remote Sensing and GIS.* New York: Lewis Publishers.

Rosswall, T., G. Woodmansee, and P.G. Risser, eds.
1988   *Scale and Global Change.* New York: John Wiley and Sons.
Sader, S.A.
1995   Spatial characteristics of forest clearing and vegetation regrowth as detected by Landsat Thematic Mapper imagery. *Photogrammetric Engineering & Remote Sensing* 61:1145-1151.
Sample, A., ed.
1994   *Remote Sensing and GIS in Ecosystem Management.* Washington, D.C.: Island Press.
Sanderson, S.E., and C.S. Holling
1996   The dynamics of (dis)harmony in human and ecological systems. In *Rights to Nature: Ecological, Economic, Cultural and Political Principles of Institutions for the Environment*, S. Hanna, C. Folke, Karl-G. Mäler, and A. Jansson, eds. Washington, D.C.: Island Press.
Sanderson, S.E., and L. Pritchard, Jr.
1993   The Human Dynamics of Land-Use and Land-Cover Change in Comparative Perspective. Paper prepared for the IGBP/HDP Core Project Planning Committee Meeting of the Global Land-Use and Cover Change Project (LUCC), New York City, July 29-August 1, 1993.
Turner II, B.L.
1991   Thoughts on linking the physical and human sciences in the study of global environmental change. *Research and Exploration* 7:133-135.
1997a   The sustainability principle in global agendas: implications for understanding land-use/cover change. *The Geographic Journal* 163:133-140.
1997b   Spirals, bridges, and tunnels: Engaging human-environment perspectives in geography. *Ecumene* 4:196-217.
1997c   Socializing the pixel in LUCC. *LUCC Newsletter* No. 1(Feb,):10-11.
in   Frontier of Exploration: Remote Sensing and Social Science Research. *Proceedings of*
press   *Pecora 13.* Bethesda, Md.: American Society for Photogrammetry and Remote Sensing.
Turner II, B.L., D. Skole, S. Sanderson, G. Fischer, L. Fresco, and R. Leemans, and others
1995   Land-use and land-cover change: science/research plan. *IGBP Report No. 35/HDP Report No. 7.* Stockholm and Geneva: International Congress of Scientific Unions and International Social Science Council.
Turner, M.
1990   Spatial and temporal analysis of landscape patterns. *Landscape Ecology* 4:21-30.
Turner, M., G.J. Arthaud, R.T. Engstrom, S.J. Hejl, J. Liu, S. Loeb, and K. McKelvey
1995   Usefulness of spatially explicit population models in land management. *Ecological Applications* 5:12-16.
Usher, M.B.
1981   Modelling ecological succession, with particular reference to Markovian models. *Vegetatio* 46:11-18.
Walsh, S.J., P.H. Page, and W.M. Gesler
1997   Normative models and healthcare planning: Network-based simulations within geographic information system environment. *Health Services Research* 32:243-260.
Wear, D.N., M.G. Turner, and R.O. Flamm
1996   Ecosystem management in a multi-ownership setting. *Ecological Applications* 6:1173-1188.
Woodcock, C.E., and A.H. Strahler
1987   The factor of scale in remote sensing. *Remote Sensing of Environment* 21:311-332.

4

Linking Satellite, Census, and Survey Data to Study Deforestation in the Brazilian Amazon

Charles H. Wood and David Skole

Advances in remote sensing technology undoubtedly rank among the most significant contributions to the study of environmental topics in recent decades. The ability to use orbiting platforms to measure the magnitude, pace, and pattern of land-cover change has been particularly relevant to the study of the Brazilian Amazon, a region that has experienced one of the highest rates of deforestation worldwide.

High-resolution satellite data provide a firm empirical base for measuring the amount and the spatial configuration of forest clearing, but they do not themselves explain the causes of deforestation. It is well understood that, beyond the need for refined measurement, explanations and projections of land-cover change depend critically on the ability to model the social determinants of deforestation. When the concern is for large regions, such as the Amazon Basin, population and agricultural censuses are virtually the only source of regionwide data on the socioeconomic and demographic characteristics of the population. These considerations suggest the prospect of modeling the causes of deforestation by using a data set that links the estimates of land-cover change derived by satellite images to the social indicators generated by the various censuses.

A regionwide research design based on satellite and census data was a natural vehicle for a productive collaboration between a systems ecologist with expertise in remote sensing technologies (Skole), and a social demographer with field experience in the Brazilian Amazon (Wood). Collaborations of this kind, although hardly new, have been relatively rare, at least in the context of Amazonian research. In the case of this particular collaboration, the joint effort can be traced in large measure to trends internal to both research traditions, the implications of which provided the impetus for the present partnership.

The National Aeronautics and Space Administration (NASA)-funded Pathfinder project achieved international recognition for its singular contribution to the production of accurate estimates of deforestation for the Amazon region as a whole (Skole and Tucker, 1993). The estimates, which were years in the making, represented a timely contribution to a controversial and highly politicized debate regarding the amount of land clearing that had taken place in northern Brazil (e.g., World Bank, 1992). Although disputes of this import are never fully resolved, it is safe to say that the publication of the results went a long way toward settling some of the major controversies in the field.

Ironically, perhaps, the findings also underscored a fundamental limitation of the Pathfinder data on deforestation, namely the inability to explain the reasons for the observed outcomes in land-cover change. Concern over the limited ability to explain the social causes of deforestation has grown in recent years in proportion to the priority accorded the so-called human dimensions of global change. Attention to these human dimensions within both scholarly and funding institutions, in turn, has compelled members of the remote sensing community to go beyond the question of "how much?" to address the question of "why?" The change, sometimes stated in terms of a shift in focus from "pattern to process" (Skole, 1997), means remote sensors have increasingly been thrust from the relative safety of the grid-cell maps to which they were accustomed into the turbulent waters of economics, politics, and sociology and other disciplines within the social sciences.

Wood arrived at the partnership by traveling in the opposite direction. After completing a 15-year longitudinal study of a particular site within the Brazilian Amazon, he had grown impatient with the "why?" question and wanted, instead, to know "how much?" The in-depth study of frontier change in which he had been engaged produced a detailed account of the events that led to the massive deforestation of the southern region of the state of Pará (Schmink and Wood, 1992). Yet for all the advantages of the "thick description" produced by the case study method, the findings were inherently limited to a single locale, leaving unanswered whether the same degree of deforestation was under way in other parts of the region. From the vantage point of the research site, it was impossible to determine whether southern Pará was a special case or was typical of what was happening elsewhere.

Despite the disparate routes taken to arrive at the point of collaboration, it was easy to agree on the main objective of the project—to explore the feasibility of constructing a regionwide model of the determinants of land-cover change in the Brazilian Amazon. The modeling exercise has the potential to address two different albeit related lines of inquiry. One, which is common to the questions raised by social scientists, looks to empirical results for explanations of the deforestation in the region. Attention focuses on the covariates of land-cover change as a means to identify and rank in importance the socioeconomic and demographic variables associated with forest clearing. Another line of research, more common to global modelers, looks to the statistical covariations as a tool for projecting the probable future levels of deforestation that are likely to be associated with assumed changes in the socioeconomic indicators.[1]

At this stage in the project, it remains to be seen whether the joining of satellite and census data can produce robust and valuable findings. It is worth learning this for the simple reason that it is always more cost-effective to use existing sources of data than to produce new data. Since both satellite images of land-cover change and census-based indicators of sociodemographic change are available for many parts of the world, the lessons learned from this effort are potentially applicable to places beyond the Brazilian Amazon.[2] Even if our objective is met only partially, a careful assessment of the strengths and the weaknesses of the research design can provide important insights for similar projects in other regions of the world.

With such an assessment in mind, our purpose here is to summarize the methods and present the preliminary results of our NASA-funded project. To establish the substantive context for the discussion, the next section defines the geographic scope of the study and presents a brief history of the factors that led to the migration of people into Amazonia and to the clearing of vast stretches of tropical forest. Next we summarize the rationale for using satellite and census data to construct regionwide models of the social and demographic determinants of deforestation. The following section describes the methods used to estimate land-cover change and to generate the sociodemographic indicators. We then review the problems associated with merging the two types of data. The final two sections present the findings of our initial efforts to establish the covariates of deforestation in the Amazon and describe a proposed method for using field work to establish ground truth for the statistical models.

THE BRAZILIAN AMAZON

Defining the Region

The geographic boundaries of Amazonia can be defined in various ways. The Amazon River drainage basin in the South American continent is an area of approximately 6,600,000 km^2 that includes land in Brazil, Colombia, Ecuador, Peru, Bolivia, and Venezuela. Within Brazil, the states of Acre, Amapá, Amazonas, Pará, Rondônia, and Roraima—an area of around 3,500,000 km^2—are referred to as "Classical Amazonia" or "North Region." The last and most commonly used definition (and the one used here) is the "Legal Amazon," a federal planning designation that conforms more or less to the watershed within Brazil's national boundaries. It consists of the North Region, plus the states of Mato Grosso, Tocantins, and Maranhão west of the 44th Meridian (Figure 4-1).

FIGURE 4-1   States of the Legal Amazon, 1980.

Development Policy, Land Settlement, and Deforestation

The contemporary movement of people into the Brazilian Amazon began in the 1970s when the agricultural frontier moved into the northern states of Pará, Tocantins, and Rondônia. Whereas earlier periods of expansion were relatively spontaneous, the exploitation and settlement of Amazonia in the 1970s were aggressively promoted by the federal government. Development policies designed to populate the region included credit and tax incentives to attract private capital to the region, construction of the Transamazon Highway, and the colonization of small farmers on 100-hectare plots along both sides of the new road (Fearnside, 1986; Moran, 1981; Smith, 1982). Colonization projects organized by the Institute of Colonization and Agrarian Reform (INCRA) attracted migrants from all parts of Brazil, who soon arrived in numbers that far exceeded INCRA's capacity to absorb in the planned communities. With few alternatives available to them, newcomers to the region who could not find a place in the colonization areas simply cleared whatever land they could find, often to be dispossessed later by ranchers and land speculators (Wood and Schmink, 1978). In as little as 2 or 3 years, places that once held a handful of families exploded into makeshift towns of 15,000 to 20,000 people. According to recent estimates of population growth in the Amazon, the states in northern Brazil experienced a net in-migration of nearly 1.6 million people between 1970 and 1991 (Wood and Perz, 1996).

At the same time that small farmers migrated northward in search of land, well-financed investors took advantage of profitable tax and credit programs offered by the Superintendency for the Development of the Amazon (SUDAM) to convert huge tracts of land to pasture and to buy land to hold in investment portfolios as a hedge against future inflation (Hecht, 1985; Mahar, 1979). Capitalized investors came mostly from the southern part of the country, where land values were high relative to the price of land in the Amazon. In the early 1980s, for example, a rancher could obtain 15 hectares in the Amazon for every hectare sold in the south (World Bank, 1992:12-13). To increase the size of their holdings substantially, ranchers sold out in the south and moved to the north, where they cleared the forest for pasture. The tendency to deforest among large landholders was further stimulated by the progressive features of Brazil's Land Statute, which levied a 3.5 percent tax on the value of unused (i.e., forested) land (Binswanger, 1991).

Although much, if not most, of the deforestation that took place in the Amazon was carried out by medium- and large-scale ranchers, small farmers were also implicated in the process, as evidenced by the typical cycle of land use. Small farmers commonly clear 2 to 3 hectares of land, which they cultivate for as long as soil fertility remains high. In most areas, soil fertility is depleted in 2 to 3 years, necessitating the clearing of more land. Since there are approximately 500,000 small farmers in the region, these figures imply a demand for an additional 500,000 hectares of cleared land per year (Homma et al., 1992:9). Crude as these estimates may be, they nonetheless point to the magnitude of the existing internal demand for land clearing, even if the migration of small farmers to the Amazon were to stop altogether.

Beginning in the mid-1970s, violence became commonplace as cattle ranchers, land speculators, peasant farmers, and Indian groups competed for control of the newly accessible territories. In a rural context characterized by violent competition for land and in the absence of clear property rights to guarantee ownership, individuals asserted their land claims by clearing the forest cover, often to a much greater degree than was economically necessary.

The direct cause of deforestation in Amazonia was thus the change in land use that came about as a consequence of the decline of fishing, forest extraction, and shifting small-plot agriculture. These traditional forms of rural sustenance were replaced in economic importance by the emergence of large peasant farming communities and the creation of pastures for cattle raising associated with the influx of migrants into the region. Table 4-1 presents estimates of the magnitude

TABLE 4-1 Area Deforested in Legal Amazon,
Brazil, 1978 and 1988

	Area Deforested (in km^2)	
State	1978	1988
Acre	2,612	6,369
Amapá	182	210
Amazonas	2,300	11,813
Maranhão	9,426	31,952
Mato Grosso	21,134	47,568
Pará	30,449	95,075
Rondônia	6,281	23,998
Roraima	196	1,908
Tocantins	5,688	11,431
Total	78,268	230,324

SOURCE: Skole and Tucker (1993:1906).

of deforestation in various states in the Legal Amazon during 1978 and 1986. The results indicate that the size of deforested areas rose from 78,268 km^2 in 1978 to 230,324 km^2 in 1988. These figures imply an annual average rate of deforestation of 15,000 km^2 per year during the period. The table shows that the highest rates of land-cover change in both years took place in Pará and Rondônia, the primary destination of heaviest migration flows into the region.

RATIONALE FOR THE RESEARCH DESIGN

Analysts have applied different approaches to study the determinants of deforestation. Numerous cross-country studies conclude that population growth and land-cover change are strongly correlated (e.g., Allen and Barnes, 1985; Rudel, 1989). Studies of this kind, however, based on highly aggregated units of analysis (countries), generally offer limited insights into the dynamics of land-cover change as compared, for example, with regional analyses that are carried out within countries and make use of a wider range of independent variables (e.g., Reis and Guzmán, 1992; Pfaff, 1997). The most detailed results unsurprisingly come from case studies, which have been especially valuable in producing highly nuanced analyses of particular sites in the Brazilian Amazon. Examples include studies of the history of land settlement in southern Pará (Schmink and Wood, 1992), surveys of the agricultural practices of colonists in the Altamira region (Moran, 1981; Walker et al., 1993), and economic assessments of public and private colonization projects (Almeida, 1992). Most studies carried out in this tradition have relied on interviews and surveys, although more recent analyses have sought to combine the data produced by conventional social science methods with the satellite-generated information on a particular scene (e.g., Brondizio, 1996; Moran et al., 1994a; Moran and Brondizio, in this volume). The case study approach has been especially valuable in producing detailed, often historically informed treatments of the events that are the cause of land-use and land-cover change in a particular place.

Yet for all its advantages, the time and resource intensity of the case study method precludes its application to large areas. Moreover, the conclusions generated by a case study approach do not go very far toward answering a broad range of questions. For example, the availability of satellite images for large areas—in this case, the data for the entire Legal Amazon produced by the NASA Pathfinder project—opens up the possibility of constructing regionwide models of the human dimensions of deforestation. Regional models, in turn, respond to the call for empirical results that are comparable from one country to another and can serve as inputs to improve the modeling and projection of various kinds of global dynamics. Many of the global change models, including those dealing with climate and trace-gas dynamics, rely on projections of land-cover change for countries across the world.

The latter projections require a coordinated program of comparative studies conducted at the regional level that specify the relationships between land-use and land-cover change, and a common set of independent variables (and their surrogates), such as changes in population size, distribution, and density, and changes in economic structure and technology. With these objectives in mind, it is worth noting that census data—not only in Brazil, but in other countries as well—are nearly always the only source of comparable sociodemographic data for large areas. By the same token, satellite images, which can be obtained for almost any place on the globe, are virtually the only source of accurate and georeferenced data on land cover for large geographic expanses.

Because of their potential contribution to global environmental modeling, the production and testing of regional models is a goal that has been promoted by a host of influential international institutions. The International Geosphere-Biosphere Programme (IGBP) and the International Human Dimensions Programme on Global Environmental Change (IHDP), through the Land-Use/Cover Change (LUCC) project, have called for a common protocol for studies that make use of existing data sources (Turner et al., 1994:93). The goals set forth in the IGBP-IHDP agenda are echoed in parallel documents produced by the Committee on the Human Dimensions of Global Change of the International Social Science Council (ISSC), prepared in cooperation with UNESCO (Jacobson and Price, 1991), and by the National Research Council (NRC) (1992). Similar recommendations have been put forth by the Social Science Research Council Committee for Research on Global Environmental Change and the 1991 Global Change Institute on Global Land Use Change, sponsored by the Office of Interdisciplinary Earth Studies. A data set that merges satellite-based estimates of land-cover change with census-based indicators of socioeconomic and demographic structure thus has the potential to go a long way toward meeting the aims of the IGBP-ISSC-NRC scientific agenda.

DATA AND DESIGN

Satellite Estimates of Deforestation

The measures of deforestation used in this study were generated by Skole as part of the Landsat Pathfinder Tropical Deforestation Project (funded by NASA, the Environmental Protection Agency, and the U.S. Geological Survey) at the Institute for the Study of Earth, Oceans and Space of the University of New Hampshire, in collaboration with NASA's Goddard Space Flight Center and the Department of Geography at the University of Maryland. Images were obtained from the U.S. national archive at the Earth Resources Observation System (EROS) Data Center, from foreign ground stations, and from programmed acquisitions.

The satellite data were preprocessed at the EROS Data Center to a standard format and projection (Universal Transverse Mercator) and sent on 8 mm tape to the University of New Hampshire for analysis. The image thresholding method was used to identify seven thematic features: forest, deforestation, secondary forests, water, clouds, cloud shadows, and cerrado (natural savanna). The databases were compiled from 210 Landsat Multispectral Scanner (MSS) scenes at a spatial resolution of 57 m. Because of both the methodology used and the nature of digital remote sensing, the output classification was not entirely accurate. Therefore, the classification was edited manually using the geographic information system (GIS). The vector product was plotted at 1:250,000 scale on vellum using an electrostatic plotter. The vellum plot was then overlaid on a 1:250,000 scale colorfire photoproduct of the Landsat scene, and misclassified polygons were identified and corrected. The vector coverage was repeatedly plotted and checked until the classification had been completed. Individual digitized scenes were projected into geographic coordinates (latitude and longitude), edge-matched, and merged into a sinusoidal equal-area projection to create a final digital map from which all calculations of area were made.

The areas of the Amazon deforested by human activities were defined using spectral characteristics of deforested sites. These characteristics were developed through field measurements at five calibration sites in the basin. It is rather easy to distinguish deforested areas from intact virgin forests since the spectral characteristics of the two are very different. Because there are some problems in differentiating cerrado, the study was confined to the closed-canopy forest region. Accuracy assessment was performed in the field using the Global Positioning System (GPS) and standard methods of assessment based on contingency tables. Overall accuracy was better than 95 percent for more than 300 checkpoints. Kappa and Tau statistics were also computed following the method of Ma and Redmond (1995) (97 and 99 percent, respectively). In addition, sample SPOT scenes at 20 m resolution were compared with our analysis, which used Landsat MSS and Thematic Mapper (TM) data. These intersensor comparisons were in agreement to within 6 percent. A complete description of the methods used to process and analyze the deforestation data set can be found in Skole (1992) and on the project Web site (http://pathfinder-www.sr.unh.edu/pathfinder).

Cloud cover is a serious problem in the tropics. However, using the entire catalog of all Landsat scene acquisitions (several hundred thousand) contained in the U.S. and Brazilian archives made it possible to select a data set for specific years that was generally free of clouds. The data set reported here was almost completely cloud free; less than 10 percent of the surface area was contaminated by clouds, mostly in the state of Amapá (Skole and Tucker, 1993).

A regional portrait of the spatial distribution of deforestation is given in Plate 4-1 (after page 150), for circa 1986. Identical estimates are forthcoming for circa 1992. The pattern of deforestation in 1986 clearly shows a crescent shape that corresponds to the expansion of the agricultural frontier into the southern part of the Amazon region. The land-cover change associated with the construction of roads is similarly evident in the lines of deforestation that stretch across the center of the map. The presence of large areas of savanna is also evident, as indicated by the band across the bottom of the map and several patches to the far north. It is important to eliminate the savanna regions in analyses of deforestation because these areas are not the outcome of deforestation, but were always naturally unforested.

Census Estimates of Demographic and Economic Structure

Indicators of demographic and economic structure presented here were derived from the 1980 population and agricultural censuses. Identical indicators are forthcoming for 1991 (the date of the most recent demographic census). For the demographic estimates, we used a micro data set that represented 25 percent of the complete enumeration. The large sample size enabled us to disaggregate the variables down to the municipio level. Because the data are available in the form of individual records, we were able to generate a number of indicators that are not present in the published materials. The census tapes, for example, contain information on 86 variables, of which 26 refer to housing characteristics and the remaining 60 to characteristics of individuals (with personal identifiers removed). The latter variables include information on age, sex, relationship to head of household, rural-urban location, place of birth, migration, length of residence, education, marital status, occupation, industry, class of worker, and income.

The data on occupation and industry categories serve as indicators of the proportion of the labor force engaged in various economic activities. Of special significance in the frontier setting are the numbers of people working on ranches and in agriculture. To these demographic characteristics we added additional variables drawn from the agricultural census, such as the number of cattle in a municipio and the area of land devoted to ranching and the production of subsistence crops, such as rice and beans. Annex 4-1 presents a list of the satellite- and census-based indicators generated for each municipio in Brazil's Legal Amazon region.

The data set for this study will thus be constructed by merging in a GIS the satellite- and census-based variables for each of the municipios that comprise the Legal Amazon (353 in 1980, 482 in 1991). The municipio boundaries in 1980 are shown in Figure 4-2. A glance at this map is sufficient to appreciate the highly irregular character of the municipios in the region. In the eastern region (in the state of Pará and in the states of Maranhão and Tocantins), population density is high, and the municipios are small in size. This pattern contrasts with the western and northern regions (especially the states of Amazonas, Roraima, and Amapá), where population density is low, and the municipios are quite large. The irregular shape of the geopolitical boundaries has important implications for the spatial correspondence between the satellite and census data.

SPATIAL SCALES AND SPATIAL CORRESPONDENCE

To merge the satellite and census data into a single data set, the land-cover data at 57 m resolution were aggregated to conform to the boundaries of each municipio, the smallest spatial unit for which economic and demographic data are available. In effect, this meant we had to reconfigure the detailed information depicted in Plate 4-1 to conform to the much larger and highly irregular geopolitical boundaries depicted in Figure 4-2, which resulted in the pattern shown in Figure 4-3. When the land-cover data were reconfigured to municipal boundaries, the crescent shape of the agricultural frontier remained plainly visible, yet transforming the data to a coarser scale was done at the cost of spatial precision.

The implications of reconfiguring the finely graded raster data to the coarser scale of municipio boundaries can be appreciated by contrasting the present data set with the ideal case. In the best of worlds, there would be a perfect correspondence between the spatial definition of the dependent and independent variables used in the analysis. In other words, the classification of land cover derived from the satellite images (dependent) would refer precisely to the land-cover characteristics of the area lying within the boundaries of each rural establishment. By the same token, the sociodemographic variables generated from the census data (independent) would refer precisely to the characteristics of the actor(s) responsible for making land-use decisions within the corresponding spatial unit. Such congruence would ensure that the dependent/independent variables were linked at the level of the decision unit involved. In this way, the analysis would maximally exploit the fine tuning made possible by advances in the production and analysis of satellite images and, by virtue of corresponding to the behavioral unit involved in land-use decisions, would avoid problems of interpretation associated with "ecological correlations" (described below).

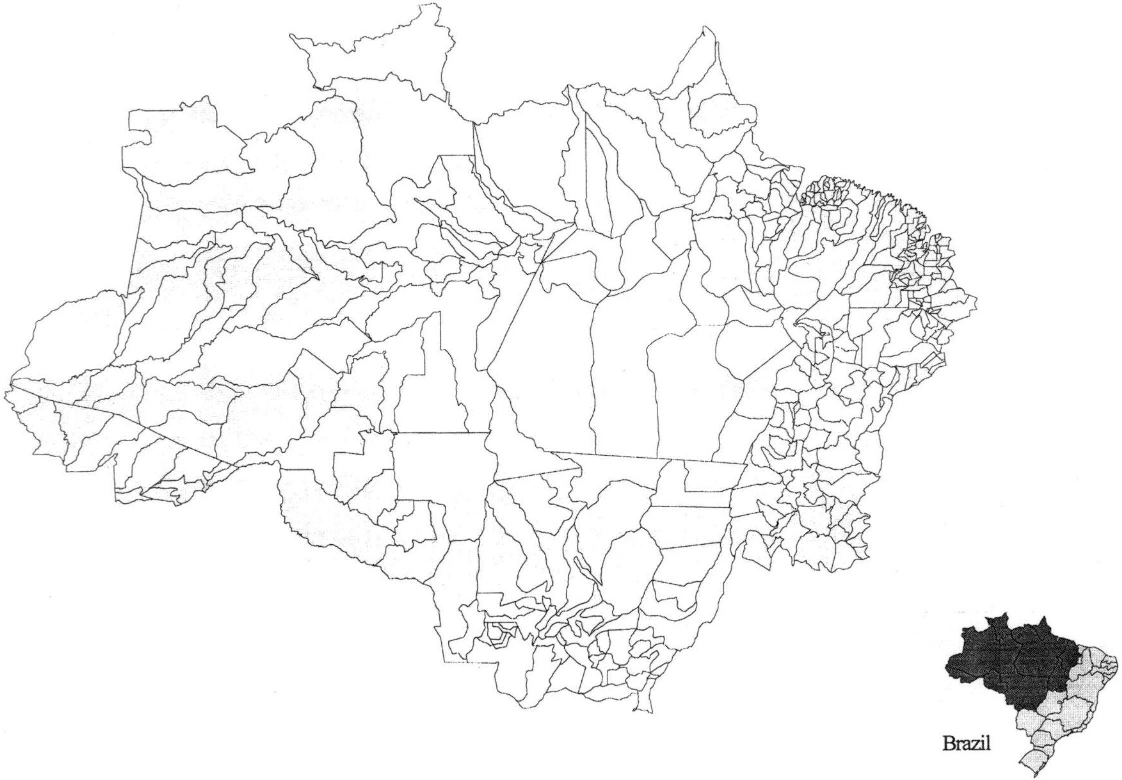

FIGURE 4-2   Municipios of the Legal Amazon, 1980.

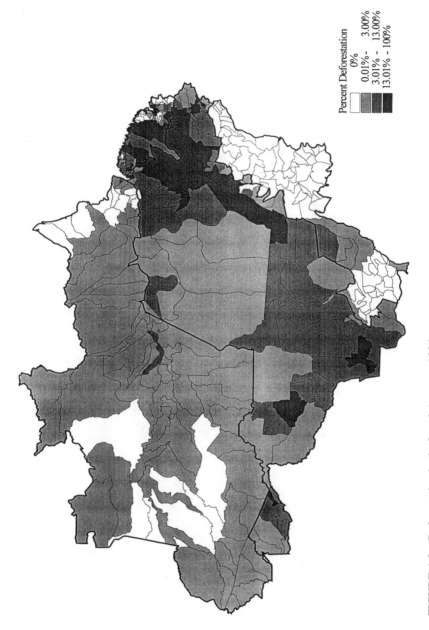

Percent Deforestation

0%

0.01% -   3.00%

3.01% - 13.00%

13.01% - 100%

FIGURE 4-3   Deforestation in the Legal Amazon, 1986.

The data at hand, however, retreat from the ideal set in two significant ways. First, when the land-cover estimates are aggregated, the values in the analysis become rates that correspond to the municipio as a whole. The result is a loss of information since the coarser scale, by reducing heterogeneity to average values, may obscure the variability of units and processes evident at a finer scale. Second, when the measures of statistical association are calculated across municipios, the data do not correspond at the level of the decision unit. Rather, there is a correlation of rates across municipios in the region. In contrast to the ideal case, in which one can be sure that the correlation reflects an association within the rural establishment, the most one can say with confidence about a correlation among rates is that the two indicators covary across units in space (in this case, across municipios). To interpret the correlations otherwise is to commit what is sometimes called the "ecological fallacy."[4] In other words, the discontinuity introduced by the fact that the dependent and independent variables are not linked at the level of the decision unit necessitates important caveats in the interpretation of the empirical relationships observed in the data.

PRELIMINARY FINDINGS

Although we have not completed the process of constructing the merged satellite/census data set, it is possible at this time to present preliminary empirical findings as a means of illustrating the kinds of analyses we have in mind.[3] The models are limited to a handful of variables, yet they point to some of the key sociodemographic determinants of deforestation in Amazonia. The variables used in the analyses are described in Table 4-2.

Given the nature of the dependent variable, the models of deforestation must account for two important features. One concerns the proportion of the municipio that is under clouds or shadows (which averaged 2.29 percent across the region). Another is the proportion of the municipio classified as savanna (cerrado). The latter is a relevant control variable because, as noted earlier, savanna areas are naturally unforested and therefore not subject to deforestation. Model 1, presented in Table 4-3, shows that although the percent under clouds is not statistically significant, the percent of the municipio considered savanna is statistically significant.[5] As expected, the sign of SAVANPCT is negative, indicating a lower level of deforestation in municipios in which a high proportion of land is savanna.

Model 1 also includes a single demographic variable—population density. The results indicate that although the relationship is statistically significant, the model has low explanatory power (R^2 = .192). Model 2 further examines the relationship between population density and deforestation by introducing the square of the density term. This specification tests the plausible hypothesis that an additional person in an already populated area will have less of a marginal impact on land clearing than an additional person in a sparsely populated area.

TABLE 4-2 Variable Names, Descriptions, and Definitions

Name	Description	Definition
AREA	Size of municipio	Total number of square kilometers within the geographic boundaries of the municipio.
DEF	Deforestation	Number of kilometers within the municipio classified as 'deforested' in 1986. Excludes naturally unforested areas (cerrado).
DEFPCT	Percent deforested	DEF/AREA.
TOTPOP	Total population	Total number of people enumerated in the municipio in the 1980 demographic census.
POPDNS	Population density	TOTPOP/AREA.
CLOUDPCT	Percent of area under clouds	Percent of the total municipio that was under cloud cover at the time of the satellite image.
SAVANPCT	Percent in savanna	Percent of the total municipio classified as naturally unforested areas (cerrado).
RMIGDNS	Rural migration density	Total number of migrants in rural areas/AREA.
RMIGSQR	Rural migration density squared	RMIGDNS squared.
FARMDNS	Farm density	Total number of heads of household classified as farmers in the 1980 demographic census/AREA.
RANCHDNS	Ranch density	Total number of heads of household classified as ranchers in the 1980 demographic census/AREA.
LT50HA	Less than 50 hectares	Percent of rural establishments less than 50 hectares in size.
GT1000HA	More than 1,000 hectares	Percent of rural establishments greater than 1,000 hectares in size.
CONFLICT	Conflict proxy	Density of cattle times density of area devoted to foodcrops.

NOTE: AREA, DEF, DEFPCT, CLOUDPCT, and SAVANPCT are from satellite images; the remainder are from demographic and agricultural censuses; see Annex 4.1.

The positive sign on the population density variable and the negative sign on its square are findings consistent with this expectation.

Total population density is, however, a variable that is subject to potential bias because the total number persons (numerator in the density ratio) includes people living in both urban and rural areas. Although the size of the urban population is not irrelevant to the study of land-cover change, it is plausible to argue that land-cover change is more closely related to the number of new arrivals in rural places. With this notion in mind, we selected the density of migrants in a rural area. We did so on the assumption that a relatively large number of

TABLE 4-3 Various Measures of Population Regressed on Percent
Deforestation (unstandardized ordinary least squares coefficients)

Variable	Model 1	Model 2	Model 3	Model 4	Model 5	Model 6
CLOUDPCT	−.302	−.191	−.063	.009	.001	.005
SAVANPCT	−.120*	−.098*	−.075*	−.108*	−.113*	−.118*
POPDNS	.311*	.609*				
POPSQR		−.002*				
RMIGDNS			8.586*	8.644*	8.594*	7.206*
RMIGSQR			−.318*	−.318*	−.317	−.348*
RANCHDNS				31.09*	30.68*	27.22
FARMDNS				4.650	4.435	2.831
FISHDNS				−11.96	−14.65	−6.149
MINEDNS				−12.60	−14.76	−17.492
LT50HA					.031	.013
GT1000HA					.171	.145
CONFLICT						.028*
R^2	.192	.237	.373	.387	.389	.453

*Coefficient is statistically significant ($p < .05$).

newcomers would correspond to areas undergoing an expansion in the agricultural frontier, which would therefore be more likely to experience a high rate of deforestation. This hypothesis is borne out by the results of Model 3, which shows that the rural migration density variable is statistically significant, as is its square, indicating diminishing effects on deforestation with a rise in density. Moreover, the R^2 (.373) is considerably higher as compared with the model based on total population density (Model 2, $R^2 = .237$).

The next step in the analysis (Model 4) took into account the economic characteristics of the population. The latter are especially valuable in the study of land-cover change because different forms of land use have different consequences for land-cover change. A municipio that has experienced the in-migration of, say, 1,000 ranchers is apt to have a higher degree of deforestation than a place that was the destination of 1,000 farmers or fishermen. Census data on the number of ranchers, farmers, miners, and fishermen thus provide proxy measures of land use. As expected, the results indicate that the percentage of land deforested is strongly associated with the density of ranchers, but not with the density of farmers, miners, and fishermen.

Additional indicators available in the 1980 agricultural census suggest the structure of land tenure in the region. Of particular interest are the percent of rural establishments smaller than 50 hectares (ha) and the percent larger than 1,000 ha. The indicators of land tenure are of potential significance in light of the widely held conviction that the majority of deforestation in the region occurs at the hands of large landholders. However, as noted in Model 5, neither variable is statistically significant, and the inclusion of these two indicators added little to the explanatory power of the model.

Finally, numerous studies of the process of frontier expansion in the Amazon have called attention to the relationship between land conflict and deforestation. In a social context in which tenure is highly insecure, landholders tend to clear large amounts of land (often far more than they can cultivate), primarily to strengthen de facto control over their land claims. Net of the effects of other factors, a high degree of deforestation can therefore be anticipated in places characterized by a high level of conflict. Conflicts over land, in turn, occur most often between ranchers and small farmers (Schmink and Wood, 1992). This observation suggests that a proxy measure of land conflict could be obtained by including in the equation an interaction term generated by multiplying the density of cattle by the percent of land devoted to foodcrops (rice and beans, which are characteristic of peasant production). Model 4 shows that, net of the other variables in the equation, the proxy for social conflict has a statistically significant association with the percent of land deforested.

FINDING ANSWERS IN THE ERRORS

The results presented above are highly schematic, based as they are on a limited number of variables and on information presently available at only one point in time. Our goal is to expand the analysis by including additional sociodemographic indicators from the population and agricultural censuses, and by generating a merged satellite/census data set for two points in time (circa 1980 and 1991). The expanded data set will allow a cross-sectional analysis of the determinants of deforestation in both years, as well as an analysis of the changes that took place over the period.

With the inclusion of additional variables, we expect to increase the explanatory power of the statistical models well beyond what has been presented here. At the same time, statistical models of real-world processes—even when they do include a much wider range of independent variables—will always contain error terms. The errors reflect the difference between the actual amount of deforestation measured in a particular municipio and the value predicted by the least-squares regression model. The errors can be visualized by tracing the least-squares regression line through the array of deforestation values the model is intended to predict. We can anticipate that some cases will fall well above the regression line (+ outliers), indicating a much higher level of deforestation than the amount anticipated by the model, while others will fall well below the regression line (– outliers), indicating a much lower level of deforestation than the model predicts.

The positive and negative statistical outliers most likely result from the failure to include in the model one or more variables highly correlated with deforestation, or what is sometimes called "specification error." If the problem is due to specification error, the only way to shed further light on the relationship is to visit the municipios in question in order to collect additional information. The goal of field work would be to identify the factors not already included in the equation that account for much higher/lower levels of deforestation than the model predicts. In other words, the strategy we have planned is to exploit all the available sources of data to construct as robust a statistical model as possible of the sociodemographic covariates of deforestation for the municipios in the Brazilian Amazon. We will then use the model to identify a handful of extreme outlier municipios (both + and –), which will become the targets of field investigation.

Field work can also be used to address another type of potential error. In addition to the problem of model specification (which produces statistical outliers), it is necessary to pay attention to those municipios that fall along (or close to) the regression line. In such cases, we are tempted to conclude that the model "works"—that the correlations we find in the model reflect true causality. The problem with drawing such a hasty conclusion is the possibility that other unmeasured factors, correlated with the independent variable(s) in the equation, are the true causes of the relationship. In this situation, the associations produced by the statistical model are said to be "spurious." As in the case of specification error, the only way to be sure that the relationships depicted in the equation are faithful representations of real events is to visit the municipios in question in order to collect additional information.[6]

Note that the research design for the field component of the project is systematically derived from the results of the modeling exercise. Research sites will be selected by identifying a subset of municipios that are statistical outliers (to address the issue of specification error) and a subset of municipios that are not outliers (to reduce the potential for misinterpretation due to spurious associations). Similarly, the content of the questions to be asked in the field will be tailored to the different research sites: in the case of positive (negative) outliers, the purpose of the inquiry will be to determine what factors beyond those included in the statistical model account for the higher (lower) level of observed deforestation; in the case of municipios that are not outliers, the purpose will be to explore the possibility that the associations found in the statistical model may be due to other, unmeasured factors.

What we propose to do in the field work stage of the project can be thought of as a variant of conventional ground truthing. The term generally refers to a process by which one verifies the interpretation of a remotely produced image. When the analyst assumes that a given pixel is, say, a deforested area, field work is carried out to ground truth the image, making sure that the interpretation is, in fact, correct. For the most part, the task is limited to establishing the correspondence between the signature of a given pixel and what is actually observed on the ground. In this sense, ground truthing is a procedure that is arguably more straightforward than what we have in mind.

In the context of the present study, the objective of field work is to verify a relationship established in a statistical model.[7] For example, when we determine that deforestation is highly associated with some variable in a regression equation, the question becomes whether that relationship is really what one observes on the ground. In effect, we are proposing to carry out what might usefully be called "relational ground truthing." Although we have not yet put this method into practice, it would appear to be substantially more complex than ordinary ground truthing. Among other things, it is far from clear how one designs a field project to address issues such as specification error (in the case of statistical outliers) and spuriousness (in the case of municipios on or close to the regression line). Indeed, the task of developing a relational ground truthing methodology is one of the challenges we confront in the coming year. If we are successful in doing so, our results have the potential to advance the process of integrating satellite, census, and field data in the study of deforestation in the Amazon and elsewhere in the world.

ACKNOWLEDGMENT

Thanks are due to Stephen Perz for his contribution to the construction of the demographic and agricultural indicators used here and his help in the analysis of the data.

NOTES

1 When data are available at two points in time, as in this study, it is possible to assess the results by using the model at time 1 to predict the values at time 2, and then compare the projected values with what is actually observed at time 2.

2 The potential relevance of this study to other regions in the world does not rest on the assumption that the statistical patterns observed in Brazil will apply to other places, which is unlikely. Instead, the relevance lies in testing the feasibility of merging census and satellite data in the study of the social determinants of deforestation. The findings have the potential to generate insights and caveats valuable to others wishing to apply the same or similar methods in other locations.

3 We eliminated municipios that are state capital cities, which are urban centers not relevant to the present analysis.

4 The ecological fallacy can be thought of as a special case of spuriousness in which the relationships found in the regression analyses are due to a shared spatial location, rather than a causal connection.

5 Future analyses of these data will account for spatial effects, which are important in two instances: (1) when the processes under study are intrinsically spatial, e.g., when they follow a spatial diffusion pattern or incorporate adjacency effects; and (2) when models are estimated using spatial (i.e., geographic) data for which the scale and unit of observation do not necessarily match the scale and unit of the process. In Anselin (1988), these two types of spatial effects are referred to as *substantive* spatial dependence and *nuisance* spatial dependence, respectively. Both are relevant to regression models of deforestation processes. On the one hand, substantive spatial dependence

allows explicit consideration of the effects of adjacency in the model. That is, it provides a way to model both forest dynamics and socioeconomic change as spatial (and/or space-time) processes. On the other hand, given the nature of the data used to estimate deforestation models—for example, census variables collected at an administrative unit level and indicators of forest dynamics aggregated to these administrative units—it is highly unrealistic to assume that the scale of the observational units matches that of the processes under consideration. In both instances, ignoring the spatial nature of the dependence causes problems of model misspecification.

6 Field work can also address the possibility of the ecological fallacy noted earlier.

7 Similar efforts to go beyond traditional ground truthing through extensive field work include those that attempt to develop new age classes of secondary growth (Moran et al., 1994b) and to understand management practices and intensification (Brondizio, 1996).

REFERENCES

Allen, J.C., and D.F. Barnes
 1985   The causes of deforestation in developing countries. *Annals of the Association of American Geographers* 75(2):163-184.
Almeida, Ana Luiza Osorio de
 1992   *The Colonization of the Amazon.* Austin: University of Texas Press.
Anselin, L.
 1988   *Spatial Econometrics: Methods and Models.* Dordrecht, The Netherlands: Kluwer Academic.
Binswanger, H.P.
 1991   Brazilian policies that encourage deforestation in the Amazon. *World Development* 19(7):821-829.
Brondizio, E.
 1996   Land cover in the Amazon estuary: Linking the thematic mapper with botanical and historical data. *Photogrammetric Engineering and Remote Sensing* 62(Aug.):921-929.
Fearnside, P.M.
 1986   *Human Carrying Capacity of the Brazilian Rainforest.* New York: Columbia University Press.
Hecht, S.B.
 1985   Environment, development and politics: Capital accumulation and the livestock sector in Eastern Amazonia. *World Development* 13(6):663-684.
Homma, A.K.O., R.T. Walker, F.N. Scatena, A.J. de Conto, R. de A. Carvalho, A.C.P.N. da Rocha, C.A.P. Ferreira, and A.I.M. dos Santos
 1992   A Dinâmica dos Desmatamentos e das Queimadas na Amazônia: Uma Análise Microeconômica. Unpublished manuscript, EMBRAPA, Belém, Pará, Brazil.
Jacobson, H.K., and M.F. Price
 1991   *A Framework for Research on the Human Dimensions of Global Environmental Change.* Paris: International Social Science Council with UNESCO.
Ma, Z., and R. Redmond
 1995   Tau coeffficients for accuracy assessments of classification and remote sensing data. *Photogrammetric Engineering and Remote Sensing* 61(4):435-439.
Mahar, D.J.
 1979   *Frontier Development Policy in Brazil: A Study of Amazonia.* New York: Praeger.
Moran, E.F.
 1981   *Developing the Amazon.* Bloomington: Indiana University Press.
Moran, E.F., E. Brondizio, and P. Mausel
 1994a   "Secondary Succession." *Research and Exploration* 10(4, Autumn): 458-466.

Moran, E. F., E. Brondizio, P. Mausel, and W. You
1994b Integrating Amazon vegetation, land-use and satellite data. *BioScience* 44(5, May):329-338.

National Research Council
1992 *Global Environmental Change: Understanding the Human Dimensions.* Committee on the Human Dimensions of Global Change. P.C. Stern, O.R. Young, and D. Druckman, eds. Washington, D.C.: National Academy Press.

Pfaff, A.
1997 Spatial Perspectives on Deforestation in the Brazilian Amazon: First Results and a Spatial Research Agenda. Paper presented in conference on Research Transformations in Environmental Economics: Policy Design in Responses to Global Change, Durham, N.C., May 5-6. Department of Economics, Columbia University.

Reis, E., and R.M. Guzmán
1992 "An econometric model of Amazon deforestation." IPEA/Rio de Janeiro, Working Paper 265.

Rudel, T.K.
1989 Population, development, and tropical deforestation: A cross-national study. *Rural Sociology* 54(3):327-338.

Schmink, M., and C.H. Wood
1992 *Contested Frontiers in Amazonia.* New York: Columbia University Press.

Skole, D.L.
1992 Measurement of deforestation in the Brazilian Amazon using satellite remote sensing. Ph.D. dissertation, University of New Hampshire.
1997 From Pattern to Process. Presentation at the Open Meeting of the Human Dimensions of Global Environmental Change Research Community, IIASA, Laxenburg, Austria, June 12-14.

Skole, D.L., and C.J. Tucker
1993 Tropical deforestation, fragmented habitat, and adversely affected habitat in the Brazilian Amazon: 1978-1988. *Science* 260:1905-1910.

Smith, N.J.H.
1982 *Rainforest Corridors: The Transamazon Colonization Scheme.* Berkeley: University of California Press.

Turner II, B.L., W.B. Meyer, and D.L. Skole
1994 Global Land-Use/Land-Cover Change: Toward an Integrated Study. *Ambio* 23(1):91-95.

Walker, R.T., A. Homma, F. Scatena, A. Conto, R. Carvalho, A. Rocha, C. Ferreira, A. Santos, and R. Oliveira
1993 Sustainable farm management in the Amazon piedmont. *Congresso Brasileiro de Economia e Sociologia Rural* 31:706-720.

Wood, C.H., and M. Schmink
1978 Blaming the victim: Small farmer production in an Amazon colonization project. *Studies in Third World Societies* 7:77-93.

Wood, C.H., and S. Perz
1996 Population and land use change in the Brazilian Amazon. Pp. 95-108 in *Population Growth and Environmental Issues,* S. Ramphal and S. Sindig, eds. Westport, Conn.: Praeger.

World Bank
1992 *Brazil: An Analysis of Environmental Problems in the Amazon.* World Bank Report No. 9104-BR. Washington, D.C.: World Bank.

ANNEX 4-1
SOCIAL INDICATORS FOR MUNCIPIOS IN THE
BRAZILIAN LEGAL AMAZON, 1980

Items 1-11 are from the 1980 demographic census; items 12-20 are from the 1980 agricultural census; items 21-22 are from satellite images, circa 1986.

1. Geographic Identifiers	Subregion
	State
	Microregion
	Municipio
	Deforestation Analysis Code
2. Base Variables	Total Population
	Rural Population
	Economically Active Population
	Total Households
	Rural Households
	Total Population Aged 5+
	Rural Population Aged 5+
3. Migration	Total Migrants
	Rural Area Migrants
	Northeast Origin Migrants
4. Labor Force Composition	Number of Farmers
	Number of Ranchers
	Number of Forest Product Extractors
	Number of Fishers
	Number of Miners
	Number of Day Laborers in Agriculture
5. Age, Sex Composition	Males and Females, from Ages 0-4 to 75+ at 4-Year Intervals
6. Child Survival	For Women Aged 15-19, 20-24, 25-29, and 30-34:
	Children Ever Born
	Children Surviving
7. Fertility	For Women Aged 15-19, ..., 45-49:
	Infants Born During the Previous Year

8. Income

For Total and Rural Heads of Households:
Income in Minimum Wages: <1, 1-<2,
2-<3, 3+

9. Housing Quality

For Total and Rural Households:
Housing Units with Mud Walls
Housing Units with Electricity

10. Literacy

For Total and Rural Populations Aged 5+:
Literate Persons

11. Agricultural Producers

Number of Owners, Renters, Tenants,
Occupants

12. Land Use

Total Number of Rural Properties
Rural Land Area Claimed in Properties
Land Area Under Different Uses:
Annual and Perennial Crops
Fallow
Natural and Cultivated Pasture
Natural and Cultivated Forest
Land Not in Use

13. Land Distribution

Number of Properties and Land Area in
Properties of <1 to 100,000+ ha
Number of Properties with <1 to 1,000+
ha of Cultivated Land

14. Agricultural Inputs

Use of Fertilizers
Number of Tractors
Value of Productive Goods
Value of Investments During Previous Year
Value of Credit During Previous Year
Value of Fuels Consumed
Amount of Various Fuels Consumed
Amount of Electricity Produced and
Consumed

15. Agricultural Outputs

Land Area, Production Yields for Annuals:
Sugar Cane
Rice
Beans
Manioc

Corn
Soybeans

Land Area, Production Yields, and
Number of Plants for Perennials:
Bananas
Rubber
Cacau
Coffee
Black Pepper

Number of Cattle
Cattle Sold or Slaughtered During
 Previous Year
Milk Production

16. Extractive Products

Açaí
Babassu Nuts
Rubber
Biomass Charcoal
Babassu Nut Charcoal
Castanha do Pará
Firewood
Timber
Palm Heart

17. Silviculture Products

Firewood
Timber
Paper Pulp
Plantation Trees: Andiroba, Cedro,
 Eucalyptus, Gmelina, Ipe, American
 Pine, Ucuubeira

18. Rural Industries

Sugar Cane Transformed:
 For Sugar (Production Volume and Value)
 For Cane Liquor (Production Volume and
  Value)
 For Molasses (Production Volume and
  Value)
 For Brown Sugar (Production Volume and
  Value)

Milk Transformed:
For Cream (Production Volume and Value)
For Doce de Leite (Production Volume
and Value)
For Butter (Production Volume and Value)
For Cheese (Production Volume and Value)

Manioc Transformed:
For Manioc Meal (Production Volume
and Value)
For Tapioca Powder (Production Volume
and Value)
For Tapioca (Production Volume and Value)

19. Other Industrial Activity
Total Industrial Establishments:
Number of Mineral Extraction
Establishments
Number of Mineral Processing
Establishments
Number of Metallurgy Establishments
Number of Logging Establishments
Number of Rubber Product Establishments
Number of Rubber Processing Establishments

20. Land Area
Square Kilometers—1980 Demographic
Census Estimate
Square Kilometers—Satellite-Based
Estimate

21. Land Cover
Land Area Under Forest
Land Area Under Savanna
Land Area Deforested
Land Area Under Secondary Growth
Land Area Under Water
Land Area Under Clouds
Land Area Under Shadow

22. Roads
Kilometers of Roads

5

Land-Use Change After Deforestation in Amazonia

Emilio F. Moran and Eduardo Brondizio

This chapter describes a project that linked traditional social science and biological field methods with remotely sensed data to further understanding of how human decisions about land use have influenced both rates of deforestation and subsequent secondary successional rates of regrowth in Amazonia. The impetus for this project was a workshop held in 1987 that introduced ecological anthropologists to remotely sensed data as a tool in addressing substantive social science questions at a regional scale. The workshop emphasized the importance of developing a partnership between social scientists and colleagues having sufficient expertise in remote sensing to solve the complex technical problems likely to be faced, and it did so without failing to note that this partnership would be best served if the social scientists developed a minimum level of proficiency in remote sensing to facilitate joint research and analysis.

Much of the promise of the new remote sensing techniques comes from expanding the areal extent of studies so that regional-scale phenomena such as land-use change can be addressed. The very advantages of small-scale studies (intimacy with informants, richness of the social network, insights into household structure) limit the ability of investigators to examine larger-scale phenomena. Remote sensing's larger spatial capabilities expand the kinds of questions that can be studied.

The published work on Amazonia in the 1970s and early 1980s spoke of devastating deforestation, desertification in the humid tropics, and wholesale conversion of tropical forest to pasture; it also made incorrect assumptions, such as 100 percent combustion of forest biomass (Lean and Warrilow, 1989; Booth, 1989). These themes, commonly expressed in studies based on remotely sensed data, did not ring true to those who formulated the project documented in this chapter. Past social science research in the area had noted farmers' complaints about the difficulties they faced from rapid regrowth of the vegetation cover following cutting and burning of forest (Moran, 1976, 1981). Secondary succession rapidly covered exposed ground and resulted in substantial land cover. Yet the large-scale work using remotely sensed data hardly mentioned secondary successional vegetation and rarely if ever suggested the significance of this vegetation to processes such as carbon sequestration, biodiversity, and land-cover dynamics.[1]

The result of these reflections was the decision to craft a set of proposals based on the same technology as that used by the remote sensing community— Landsat Thematic Mapper (TM) digital data—to understand land-use/land-cover change dynamics following deforestation, particularly the factors that might explain the differential rates of secondary succession. A grant provided by the National Science Foundation's (NSF) Cultural Anthropology Program enabled one of the authors (Moran) to become familiar with the theory and techniques of remote sensing. In fact, the Senior Scholar's Methodological Training Grant program[2] has provided support for several environmentally oriented anthropologists to acquire technical skills in other disciplines and has substantially increased the number of scholars engaging in this type of work. The following chapter by Entwistle et al. is an example of another means of linking remote sensing and social science. Following this 1-year learning period, grant support from the NSF Geography and Regional Science and Human Dimensions of Global Change programs made it possible to apply these newly acquired skills to the questions raised above.

The second author of this chapter (Brondizio), who had acquired some of these skills at Brazil's National Institute for Space Research (INPE), followed a reverse trajectory. He had familiarity with agronomy, vegetation ecology, land-use studies, and remote sensing research and undertook to learn social science methods, especially ethnographic skills, while pursuing a Ph.D. in environmental sciences.[3]

A common research question meaningful to the social and environmental sciences—what forms of land use lead to given rates of secondary successional regrowth in Amazonia—provided the epistemological basis for our collaboration. The choice of soil by a homesteader, the choice of area to be cleared, the method used for clearing, the timing of burning, the choice and sequence of crops planted, and the frequency of weeding all affect the rate at which pioneer species can colonize an area of land, the composition of that succession, and the differential survival of mature forest species. The study of secondary succession requires integration of conventional site-specific research methods in vegetation ecology and ethnographic data with more inclusive scales through remote sensing analysis of land-use and land-cover patterns (Moran et al., 1994; Brondizio et al., 1996).

This chapter presents examples from our work that illustrate the linking of remote sensing and human ecological questions in understanding land-use and land-cover change in Amazonia. The chapter does not describe specific methodology and technical details related to image processing, spectral analysis, and vegetation/soil inventory techniques, for which readers are referred to published papers written by the authors and their collaborators (Mausel et al., 1993; Moran et al., 1994; Brondizio et al., 1994, 1996; Li et al., 1994). As with other chapters in this volume, the objective here is to discuss how collaboration between the social scientist and the remote sensing expert developed, summarize the findings and insights gained by linking remotely sensed data to traditional social science field research, and explore ways of advancing this type of collaborative work in the future (e.g., Mausel et al., 1993; Moran et al., 1994, 1996; Brondizio et al., 1994, 1996; Li et al., 1994; Brondizio, 1996; Randolph et al., 1996; Tucker, 1996; Tucker et al., in press). The work discussed here brought together remote sensing, botany, environmental sciences, soil sciences, anthropology, and geography. It did so incrementally, as interest grew in the issues raised by our research among collaborators in Brazil and the United States. In other words, building on a core set of questions and the expertise of anthropology, ecology, and remote sensing, the project has expanded to address other concerns that flowed naturally from the original set of propositions. This expansion was anticipated and was integrated without difficulty. Even now, the project anticipates incorporating climatologists, zoologists, demographers, economists, and conservation biologists.

THE VALUE ADDED OF SOCIAL SCIENTISTS' INTEREST IN REMOTE SENSING ANALYSIS

Social scientists bring to the analysis of global change and its remote observation a concern about and an expertise in the behavior of people at the community and household levels and a desire to understand the human face behind the pixels (see Geoghegan et al., in this volume). When looking at a satellite image, for instance, social scientists are inclined to search for land-use patterns associated with distinctive socioeconomic and cultural differences. Consequently, they search for driving forces behind land-use differences and for land-cover classes that represent culturally and biologically meaningful differences, in contrast to, say, naming classes after a standard vegetation class. For example, the timing of credit availability for pasture or cocoa development can help determine when one might begin to see the appearance of these classes with higher frequency on a landscape, or the creation of a class called "roça," which is a mixed subsistence garden dominated by manioc and bananas. This poses a challenge in that if a culturally meaningful category is present (e.g., palm agroforestry) and is associated with important behavioral differences (i.e., particular steps in preparing these orchards through time until they reach the desired density), then an effort must be made to sample enough cases so that the phenomenon can be distinguished spectrally and classified (Brondizio et al., 1994). This may not always be possible, but it is a challenge that social scientists, or at least anthropologists, are likely to bring to the task of field work and land-cover classification. On the other hand, while a culturally meaningful category may exist, it may be so rare that one cannot possibly obtain enough observations to separate it spectrally, or it may not be different from other culturally differentiated classes that are spectrally alike. It is still a challenge, for example, to differentiate between many types of agroforestry and other mixed-crop systems given the current resolution of satellite images. Some of these problems may persist until such time as orbital satellites with a resolution of 1 m are available to researchers. The first satellites with this resolution are expected to be launched in the next year or two and, because they are launched by commercial enterprises, are expected to make data available more promptly than is the usual current practice and to customize the data to the needs of users.

Without a social scientist as part of the team, culturally important dimensions of land cover may quite possibly be overlooked by scientists who bring a nonlocal or purely remote point of view to the analysis of the data. The best example from our work is the discrimination of managed (palm agroforestry) from unmanaged floodplain forest in the Amazon estuary (Brondizio et al., 1996). Whereas these are vegetation classes with extremely similar structural characteristics and thus are commonly mapped together, managed floodplain forest has local economic significance that requires attention when one is studying an estuarine population at a regional scale. By combining traditional ethnoecological field techniques that elicited culturally meaningful categories (Açaisal) with spatial distribution considerations elicited by the satellite data, it has been possible to distinguish between these two culturally and economically distinct vegetations (i.e., managed and unmanaged floodplain forest). When the importance value[4] of the palm *Euterpe oleracea* reached 0.6, it became possible to distinguish spectrally a managed açaí palm agroforestry grove from the adjacent floodplain forest from which it had been developed by local farmers (Brondizio et al., 1994:261).

While there is no substitute for the use of traditional ethnoecological field data collection to obtain a deep understanding of native knowledge of the environment, it is possible in cases such as that described above to collect a sufficient number of observations to create spectrally differentiable classes of land cover from native categories—although success in this enterprise will rarely economize on data collection costs. What will be gained, as in most applications of remote sensing, is the ability to map at a regional scale the distribution of a land-cover class that is meaningful to a local community over a much larger landscape than is otherwise possible (Brondizio et al., 1996). The value of forecasting cereal harvests and yields of major commodities has been accepted for years in agribusiness. There is no reason, other than the more modest resources of the scientific community, why forecasting of harvests of locally valuable crops, such as manioc, bananas, or agroforestry groves marked by a dominant, cannot be undertaken. One of the important results of the study of palm agroforestry has been to show in dramatic fashion the very large areal extent of this economic activity, its economic value to the regional economy, and the achievement of this outcome with minimal loss of forest (a five-fold rise in economic value at less than a 2 percent loss of forest cover).

Even with current limitations on spatial resolution that can mask households' complex patterns of land use, anthropological and geographical understanding of the spatial distribution of sizes and locations of agricultural fields makes it possible to infer and interpret land-cover patterns that are distinguishable spectrally. This ability has been enhanced with the growing use and accuracy of technology that permits accurate location, such as Global Positioning System (GPS) devices.[5] However, the difficulties of distinguishing among coffee plantations, early secondary succession, and degraded pastures should not be taken lightly. Few analysts have tried, and even fewer have succeeded, to differentiate spectrally among types of crops, types of pastures, and types of agroforestry. The more homogenous a stand is, the more likely it is to be identified consistently with a high degree of accuracy, whereas for mixed and heterogeneous vegetation formations, such accuracy is difficult to obtain. Our own work has been able to differentiate among three distinct structural classes of secondary succession with an accuracy of 92 percent, and between managed and unmanaged floodplain forest with an accuracy of 81 percent (Mausel et al., 1993; Brondizio et al., 1996). These successes do not suggest that achieving these results was easy. On the contrary, many classes we wish to differentiate have remained elusive. Monitoring oil palm and cocoa, for example, has to date proven impossible given the very large spectral differences among their various developmental stages. We believe it should be possible to do so if sufficient resources are devoted to collecting enough observations for the distinct steps in the development of these plantations— a goal very different from ours of understanding secondary successional processes.

An excellent example of the application of remote sensing to fundamental issues in social science is a study by Behrens et al. (1994) that shows how settlement history mediates the effect of population pressure on indigenous land use. Sendentism and the market opportunities that promote it seem more important drivers of land-use intensification and tropical deforestation among contemporary native Amazonians than population growth itself. Village formation and cattle ranching are associated with greater landscape heterogeneity, but fewer woody species. Concentrating in large villages a population that has been distributed areally over the landscape can intensify deforestation, particularly when exacerbated by the development of pastures and irrigated rice cropping. The study of intensification is of fundamental interest to the understanding of human societies through time, and remote sensing is an excellent tool for observing the extent and intensity of its impact.

One of the most important contributions social scientists can make to this type of research is to help construct data collection protocols that capture the types of socioeconomic data most closely related to land-use dynamics. It is all too common for those outside the social sciences to try to explain land-cover change in terms of population growth, rather than applying the more nuanced approach needed to understand the relationship between population (growth, distribution, structure) and the environment. A current project of the authors involves investigating, at the level of the farm property, the role played by the demographic structure of each household in changing uses of the land, with a view to predicting rates of deforestation from a knowledge of household composition. There is a need to develop a protocol for the minimal data needed to support ecological and remote sensing analysis and also be meaningful to the social sciences. For example, such a protocol might include data related to production systems (types of economic importance), a calendar of activities throughout the year (land clearing, planting, weeding, harvesting, fallowing), soil and vegetation management techniques, the demographic composition of households and populations, time allocation in different production systems, land-tenure structure, and an ethnoecological classification of ecosystem components (Moran, 1993; Brondizio, 1996). For example, one should expect significant differences in the way land cover develops and changes through time as a product of, say, private or communally held forest. Current studies by our research group in seven Latin American countries are aimed at elucidating the impact of tenure and other social organizational arrangements on the composition and longevity of forest cover through time.

METHODS

Levels of Analysis and Site Selection

From the outset, we have followed a systematic approach to site selection and comparison. Taking a contrary view to that commonly held, we hypothesized that the differential rate of secondary succession would most likely be influenced by initial soil fertility, the history of land use of a deforested area, and the spatial pattern of land-use and land-cover classes. Soil fertility had seldom been related to or soil data collected in studies of forest ecology and succession (Buschbacher et al., 1988). Since tracking of age classes of secondary succession and biomass accumulation in such vegetation had not been performed in Amazonia using Landsat TM (and had been unsuccessful using the Multispectral Scanner [MSS]), we began our study by examining two locations. Each was characterized as having relatively fertile soils, so that if it were technically possible to differentiate stages of secondary succession, the change might be measurable in the relatively brief time span between 1984 and 1991 during which TM was available. For the sake of contrast in both environmental and land-use terms, we selected an upland site that was well known to one of us from earlier work (Moran, 1976, 1981) and an estuary area where we had done preliminary work (Murrieta et al., 1989). Following 2 years of research at these two sites, we worked at three other Amazonian sites, characterized by relatively nutrient-poor soil conditions and different patterns of land use.

One of our goals from the outset was to link detailed ethnographic data, species- and stand-level data, and land-use histories to the spectral analysis so that we could achieve not only a field-level understanding of changes in land cover, but also a regional analysis of land-use and land-cover change. Doing so required that we work with a large portion of the TM scene, and that our sampling design be distributed over the image in order to incorporate spatially variable phenomena, such as different kinds of settlements, different land-tenure arrangements, different types of vegetation, and different soil types. Thus, we sought to link the behavior of households in farms and settlements to regional-scale processes of land-cover change, especially secondary succession. We also wished to link these results to global carbon models and Amazon Basin models—a task that has been pursued more directly by Skole and his collaborators (e.g., Skole and Tucker, 1993).

Figure 5-1 illustrates the multiscale and multitemporal approach pursued in our studies. Our analysis begins by selecting locations that fit our fertility gradient design and have representative patterns of land use and population distribution. We also take into account data availability and the presence of colleagues with whom joint work might be undertaken that would enrich local institutions with both data and expertise. For the selected locations, we seek available cloud-free images of the study areas; depending on availability, we also try to obtain a set of images providing data intervals within which the processes of change can be observed, at least one of which is coincident with our field research. These data are then georeferenced and registered, exploratory spectral analysis is carried out in small subsets representative of different patterns of land cover, and this analysis is then used to carry out unsupervised classification of land cover over the entire study area. Details of our technical procedures have appeared in a number of publications (Mausel et al., 1993; Moran et al., 1994; Brondizio et al., 1994, 1996; Li et al., 1994).

We then proceed to the field, not merely to carry out field observations of land cover—the most common method of ground truthing—but also to interview at length land users who are identified from the initial analysis as having land-cover classes of interest for sampling and are distributed over the entire image. All of these visits to farms generate valuable information. Some are not entirely successful, either because the land has undergone transformation since the TM scene was taken or because there is error in the unsupervised classification. Detailed household surveys, with particular emphasis on the history of land use, are then undertaken. Following these surveys and a visit to the forest or secondary successional area, we request permission for the larger team to come to the property to carry out a detailed soil and vegetation inventory.

The resulting data are entered into a spreadsheet in the field, and adjustments are made in the sampling to ensure that a representative number of classes of

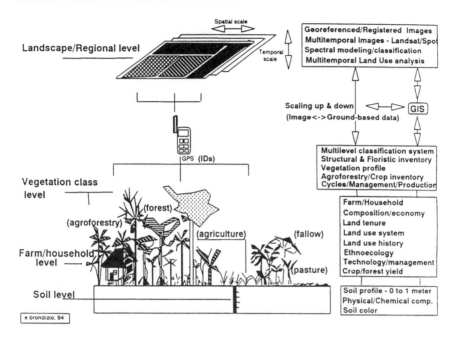

FIGURE 5-1 Methods of multilevel analysis of land-use and land-cover change.

vegetation are sampled. Each area at which soils and vegetation are sampled is georeferenced with a GPS device, every effort being made to choose study areas large enough so we can be sure of their location on the printouts of the TM scene we have prepared in advance and laminated for field use. These image printouts are generally prepared at a scale of 1:30,000 with a 1 × 1 km grid of Universal Transverse Mercator coordinates that allows us to locate each site (through use of GPS) while in the field. Fields as small as 1 hectare (ha) are clearly visible in these image maps, although we commonly select larger areas within which to take vegetation and soil samples. The use of these image printouts prepared at a fine scale and enhanced for visual interpretation has proven particularly valuable in extracting field information. The relative ease of understanding color composites (TM bands 5, 4, 3)[6] makes it possible to discuss land-use and land-cover features with local farmers with a minimum amount of explanation. Their discussion of the image provides an invaluable source of information about land-use and land-cover dynamics and makes sense out of the distribution of the different types of land cover encountered. In addition to the field-sampled plots, one or more members of the team collect "training samples" (i.e., visual observation of hundreds of locations) in order to obtain a robust supervised classification of land-cover classes upon returning from the field.[7]

Upon returning from the field, we use the GPS-referenced field observations to develop the supervised classification. We also perform accuracy analysis to determine the extent to which classification accuracy of at least 85 percent have been achieved. A second season of field work has characterized our work so far, at which time we are commonly able to double our field inventory data and refine the accuracy of the land-cover classes.

During the past 5 years our group has developed an extensive data set. This data set is focused on secondary succession and land-use and land-cover change in five Amazonian regions distributed along a soil fertility gradient representing relatively nutrient-rich (eutrophic) to relatively nutrient-poor (oligotrophic) conditions.

Study Areas

Altamira, in the Xingu Basin, is characterized by patches of nutrient-rich soils (alfisols) and less fertile soils (ultisols). Ponta de Pedras, in Marajó Island, is characterized as a transitional environment composed of upland nutrient-poor oxisols and flood plain alluvial soils. Igarapé-Açú, in the Bragantina region, is a mosaic of oxisols and ultisols. The soils of Tomé-Açu are dominated by oxisols and ultisols, both acidic and low in nutrients. They are less sandy in textural characteristics than those in Bragantina (Igarapé-Açu) but more so than those in Altamira or Marajó. Finally, Yapú, located on the Vaupés (a tributary of the Rio Negro), is composed of large patches of extremely nutrient-poor spodosols intermixed with stretches of oxisols.

Land use varies among these areas. Altamira, which lies along the Transamazon highway, began being colonized in 1971 and has experienced high rates of deforestation and secondary succession associated with the implementation of agropastoral projects. In contrast, Marajó has historically been home to native nonindigenous (i.e., Caboclo) populations occupied primarily in agroforestry activities in the floodplain and swidden agriculture in the uplands, along with some recent creation of pastures and mechanized agricultural fields. Land use in the Bragantina region has gone through several phases; today short-fallow swidden cultivation is dominant, given the proximity of the Belém market for producers. Cultivation of secondary growth areas has been common for decades, and islands of mature forest are rare. Tomé-Açú has experienced the most intensive agriculture of the five sites (a black pepper monoculture until the late 1960s), and for the past two decades has been associated with agroforestry development carried out by the Japanese colonists who have lived there since the 1930s. It is now experiencing the start of pasture formation. Finally, the Colombian Vaupés site at Yapú, populated by indigenous Amazonians, is characterized by more traditional long-fallow swidden cultivation based on bitter manioc.

1 Altamira, Xingu Basin, Pará, Brazil

2 Ponta de Pedras, Marajó Island, Pará, Brazil

3 Igarapé-Açú, Bragantina, Pará, Brazil

4 Tomé-Açú, Pará, Brazil

5 Yapú, Vaupes Basin, Colombia

FIGURE 5-2   Research sites in Amazonia of the Anthropological Center for Training and Research on Global Environmental Change (ACT), Indiana University.

Distribution of Research Locations

Figure 5-2 shows the locations of the five study areas discussed above, and Plate 5-1 (after page 150) shows Landsat images of each location. In each region, areas representative of the major vegetation types, including different forest types and fallows, are selected for sampling. Altamira is represented by 20 sites (18 fallows and two forests), Marajó by 14 sites (10 fallows and 4 forests), Bragantina by 19 sites (16 fallows and 3 forests), Tomé-Açú by 13 sites (12 fallows and 1 forest), and Yapú by 8 sites (5 fallows and 3 forests), for a total of 74 sites. The detailed soil and vegetation inventories permit careful characterization of each location.

Vegetation and Soil Inventory and Processing

Our strategy is comparable across the 74 sites. The majority of plots and subplots are identical in size and shape, allowing cross-comparison and integration at the level of plot, site, and location. In most cases (except mature forest), the area sampled per site is 1,500 m^2. Plots and subplots are randomly distributed, but nested inside each other to account for the detailed inventory of trees (diameter at breast height [DBH] greater than or equal to 10 cm), saplings (DBH 2-10 cm), seedlings (DBH less than 2 cm), and herbaceous vegetation. In the plots, all the individual trees are identified and measured for DBH, stem height (height of the first major branch), and total height. In the subplots, all individuals (saplings, seedlings, and herbaceous vegetation) are identified and counted, and diameter and total height are recorded for all individuals with DBH equal to or greater than 2 cm.

Species identification is carried out by experienced botanists in the field and checked at the herbarium in Belém, Pará. Botanical samples are collected from half of all species identified to ensure accuracy of taxonomic identification. At each site, soil samples are collected at 20-cm intervals to a depth of 1 m. Soil samples are analyzed at the soil laboratories in Belém for both chemical and physical properties. A stand inventory table, including absolute and relative frequency, density, dominance, basal area, importance value, and stem and total height, is prepared for each of the inventoried sites.

A soil fertility index summarizing differences among regions is used (Alvim, 1974). It is important to note that our comparisons take into account only upland soils of the Marajó site. The more fertile floodplain is not included in this analysis since it is not comparable at this level with the data of the other four locations, all of which have upland soils with very different characteristics from those of floodplains. The index aggregates pH, organic matter, phosphorus, potassium, calcium and magnesium, and aluminum (inverse value). It was prepared for each depth (0-20, 20-40, 40-60, and 80-100 cm), and an average index was prepared across depths.

PATTERNS OF SECONDARY SUCCESSION IN AMAZONIA

The data obtained using the methods described above have yielded a number of findings with regard to soil physical and chemical patterns in the study regions, variations in rates of regrowth, and stages of regrowth in Amazonia.

Soil Physical and Chemical Patterns in the Study Regions

Soil structure and texture in the study regions, as represented by the percentage of fine sand, coarse sand, silt, and clay at five depths (20-cm intervals) are presented in Figure 5-3. Coarse sand and clay are the elements most able to provide discrimination across regions. Four major textural groups can be distinguished in the study regions (see Figure 5-3). Altamira soils have a low content of fine and coarse sand at all depths (average around 10 percent) and high clay content (above 45 percent) at all depths. Although the Yapú region presents a similar textural pattern, it can be distinguished by the presence of a spodic-B horizon with low permeability and penetrability, characterized as a groundwater humic podzol (Sombroek, 1984). Marajó and Bragantina soils are rather similar in terms of sand and clay content at all depths. In both cases, average, fine, and coarse sand content are above 25 percent and average clay content is below 20 percent at all depths. Tomé-Açú soils, although similar to those of Marajó and Bragantina, are distinct in their lower content of fine sand (below 25 percent) and higher clay content (30 to 40 percent) at all depths. Therefore, while Marajó and Bragantina offer typical examples of oxisols, Tomé-Açú presents a soil type closer to an ultisol.

Differences in soil fertility are small but significant among the study regions. Altamira is clearly superior in terms of soil fertility, while differences among the others are minor. The average pH of above 5 in Altamira contrasts with the average pH of below 5 in the other regions. A pH of above 5.5 is viewed as necessary for crop productivity from all but a few domesticated cultigens, such as manioc, cowpeas, and sugarcane, that are adapted to low-pH conditions. However, among the regions with lower-pH, less-fertile soils, the soils of Marajó and Yapú have lower pH (below 4.4 in the first 20 cm or plow layer) than do those of Bragantina and Tomé-Açú.

Combined analysis of aluminum and calcium/magnesium content further distinguishes fertility among the five regions. Yapú is the most nutrient-limited region, with the highest aluminum concentrations and the lowest concentrations of calcium and magnesium. This pattern is reinforced by the low availability of phosphorus. High aluminum saturation tends to limit absorption of other nutrients, especially calcium and magnesium, because of impeded root development (Lathwell and Grove, 1986). Phosphorus is considered the most limited nutrient in Amazonia, frequently found only in trace quantities (below 1 part per million). It is low at all the sites, although Altamira has slightly larger amounts than the other regions. No significant differences are found among regions in the amount of organic matter. Analysis of these elements in the form of a fertility index, as shown in Figure 5-4, reveals that soil fertility is significantly different between Altamira and the other regions, but similar overall among the other four regions.

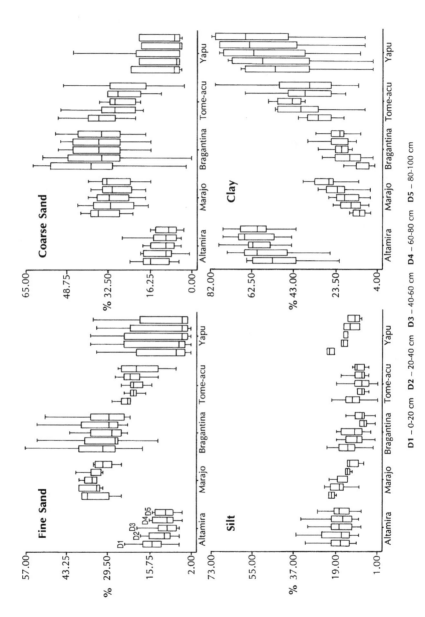

FIGURE 5-3   Soil texture by depth (Altamira, Marajó, Bragantina, Tomé-Açu, Yapú).

D1 – 0-20 cm **D2** – 20-40 cm **D3** – 40-60 cm **D4** – 60-80 cm **D5** – 80-100 cm

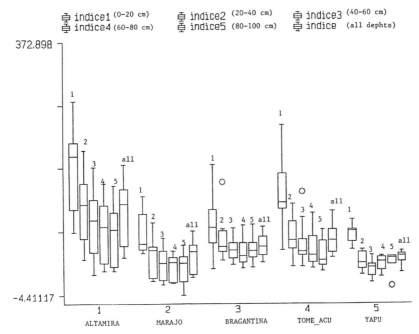

FIGURE 5-4   Soil fertility indexes for five research sites (pH + OM + P + K + Ca + Mg – Al).

Variation in Rates of Regrowth

Analysis of variance shows that soil fertility is a significant indicator of differences among regions with regard to secondary succession (adjusted r^2 = 0.69; p < .05) when average stand height is used as a parameter to indicate rates of regrowth. Differences in fertility clearly favor regrowth in Altamira and differentiate it from the other regions. Similar regrowth rates are observed in Marajó, Bragantina, Tomé-Açú, and Yapú. These differences are shown in Figure 5-5, where the average regrowth rate for Altamira is compared with the average rate for each other region. Altamira is the only region showing above-average rates of regrowth. During the first 5 and 10 years of fallow, Altamira fallows are 1 m higher as compared with the average fallow of all other regions. This difference increases two-fold after 15 years of fallow. Such an increase may be closely related to the faster development of trees in relation to saplings in this region. Overall, Altamira tends to reach higher canopy and lower understory biomass density at a faster rate than the other sites, indicating a more rapid pace of tree and forest structure development. This pattern is reinforced by the differences in family diversity between Altamira and the other regions.   Overall,

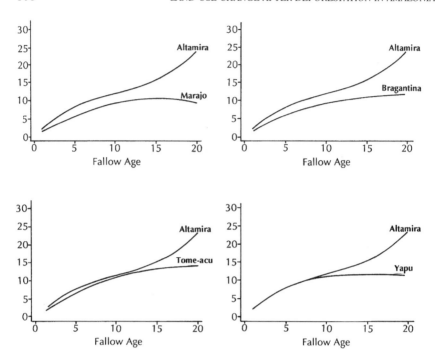

FIGURE 5-5 Height increment in secondary succession.

family diversity in the understory is higher during the first 5 years of fallow, decreases during the following 10 years, and increases again as vegetation reaches a forest-like structure. In the canopy, family diversity presents a progressive rate of increase with age. However, significant variations in this pattern can be perceived. Whereas Altamira has a lower understory diversity as compared with the other regions (especially Bragantina), it has the highest diversity of tree families, reflecting its faster canopy development. The Bragantina region in particular presents in some cases a higher degree of understory family diversity within a short-fallow swidden cultivation system. The greater diversity of saplings and herbaceous vegetation in this region is closely associated with resprouting of a specific group of families and species that has made it possible for them to survive under this intensive land-use system (Denich, 1991; Vieira et al., 1996). However, the relationships among soil fertility, fallow cycle, and indicator species are still unclear and will be the focus of attention in the near future.

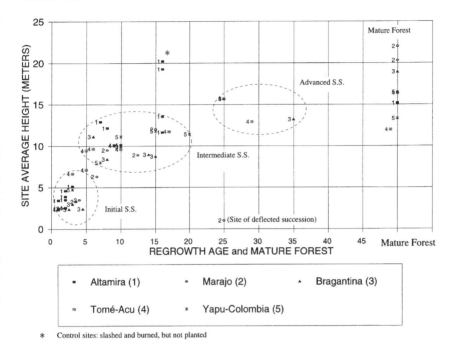

FIGURE 5-6 Regrowth stages and average stand height, ACT research sites in Amazonia.

Basin-Wide Patterns of Rates of Regrowth:
Defining Stages of Regrowth in Amazonia

An important finding of this study is the definition of basin-wide stages of regrowth based on the analysis of average stand height and basal area of the study locations that correspond to distinctions derivable from spectral analysis of satellite data. While this finding may not seem to represent an important social science contribution, the detection of rates of regrowth illustrates the use of remote sensing to discriminate social phenomena, specifically land-use patterns, thus socializing the pixel (see Geoghegan et al., in this volume). Examination of the literature supports our analysis by revealing data consistent with the characteristics of our study areas. Stand height proves to be a significant discriminator of regrowth stages at the level of the study area (Figure 5-6). Three structural stages of regrowth can be distinguished: initial (SS1), intermediate (SS2), and advanced (SS3) secondary succession. These stages can be broadly associated with age classes.

SS1 is associated with a period of establishment dominated by herbaceous and woody species. Saplings are the main structural element and represent the majority of the plant biomass. Average height is 2 to 6 m. Most individuals are of a height equal to or less than 2 m during the first 2 years of fallow. In terms of basal area, this stage presents a variation that ranges from 1 to 10 m²/ha. The majority of individuals have DBH of 2 to 5 cm. In age terms this period encompasses around the first 5 years of fallow, but it may be much longer in areas subjected to heavier land-use impacts.

SS2 can be characterized as a period of thinning of herbaceous and grass species and a rapid increase in sapling dominance, with small trees beginning to appear. While saplings account for most of the total basal area and biomass, young trees dominate the canopy structure. Canopy and understory become increasingly differentiated, but stratification is still subtle. The increase in shade during this stage is an important element in species selection. Average height is 7 to 15 m, and DBH is 5 to 15 cm. Basal area ranges from 10 to 25 m²/ha. This period encompasses the next 10 years fallow, that is, 6 to 15 years after abandonment.

SS3 is marked by a growing stratification between understory and canopy and by the declining contribution of saplings to total basal area and biomass. Average height is 13 to 17 m; however, a considerable number of shorter (6 to 13 m) and very tall (20-30 m) emergents occur. Individuals with DBH of 10 to 15 cm are still of major importance at this stage, but a considerable number of larger individuals are present. In terms of basal area, this stage is similar to the intermediate stage. One of the reasons for this relates to the process of species selection that occurs between SS2 and SS3. Fast-growth trees of SS2 (e.g., *Cecropia* spp.) that contribute a major portion of the total basal area at this stage tend to give way to forest tree species during SS3. Therefore, instead of a progressive increment in basal area from SS2 to SS3, there is replacement of the species and individuals contributing to basal area. In age terms, this is a stage that encompasses fallows of more than 15 years. However, we found that in Altamira, this type of structure could be achieved in about 11 years.

Mature forest vegetation varies widely within the Amazon. Average height varies from less than 15 m to around 24 m. However, one can distinguish between forest and advanced regrowth by taking into account additional features that characterize mature forest vegetation. First, species composition needs to be considered as a unique discriminator of a mature forest environment. Mature forests have higher species diversity. The presence of very tall emergent trees with large diameters is also a distinctive feature. Most emergent trees have DBH above 30 cm and height greater than 15 m. Basal area in mature upland forest ranges from 25 to 50 m², thus providing a distinct structural difference from advanced regrowth (10 to 25 m²/ha) that facilitates spectral separation.

This model of regrowth stages can be applied to the Amazon region if land-use intensity, landscape diversity, and soil fertility variables are taken into account at the regional and local levels. The proposed regrowth classes provide a baseline for remote sensing analysis and large-scale studies of land-use and deforestation dynamics. Structural characteristics of the vegetation such as those described above influence spectral information. Differences in average stand height can be correlated with increased absorption (i.e., lower reflectance) of the visible bands (i.e., 1, 2, 3 in Landsat TM). Likewise, these structural features lead to differing reflectance values among SS1, SS2, and SS3 and mature forest in the near- and mid-infrared bands (i.e., 4, 5, and 7). Understanding these structural/spectral relationships gives social scientists a powerful tool for studying land use and agricultural cycles of human populations. The structural parameters presented above allow one to discriminate with modest effort between areas recently and long abandoned, and to collect good-quality training samples for image-supervised classification (see Mausel et al., 1993, for a fuller discussion of spectral characteristics of vegetation types).

If one keeps in mind the general features of land-cover classes and which features have the greatest role in spectral differentiation, field research observations can generate information of considerable value to the classification of land cover. Good-quality pastures have higher visible-band reflectance than degraded pastures because of their more homogeneous surface and minimal shadow and the presence of soil as a component of reflectance (also producing a relatively high mid-infrared reflectance). A degraded pasture and initial secondary vegetation will have a lower visible-band response due to greater vegetation, less soil, and increased chlorophyll absorption. We take the categories of degraded pasture and SS1 to be equivalent: a degraded pasture is a cultural category meaningful to a rancher who sees a pasture that has been invaded by woody growth and is no longer capable of sustaining cattle; SS1 is an ecologist's category, used when a rich diversity of pioneer species occupies land that was previously cultivated or deforested and that had been characterized by a dominance of herbaceous and woody species. In degraded pasture or SS1, the near-infrared will have higher reflectance due to mesophyll reflectance, but the mid-infrared will have lower reflectance than in pasture as a result of greater absorption of water by the vegetation.

The developing canopy in SS2 has higher biomass and moisture than are found in SS1. Thus, while the visible bands will not differentiate it from SS1, the green-to-red ratio will be higher. The mid-infrared is lower than in SS1 because of increased shadow, a pattern that continues with the greater growth of the vegetation. In the advanced stages of regrowth, near-infrared and mid-infrared reflectance continues to drop because of increased shadowing and increased moisture levels in the vegetation.

Thus it is important for field work to distinguish the pattern of canopy and understory, the amount of exposed soil, the surface roughness of the vegetation, and the amount of shadowing to assist in spectral analysis of these patterns in the satellite data.

IMPLICATIONS FOR TRAINING AND RESEARCH

One of the lessons that emerges from our experience is the importance of taking time to learn the basic theoretical and methodological approaches of collaborators in other disciplines. While it may be possible to develop a clear division of labor between collaborators, the collaboration will proceed more smoothly if social scientists are familiar with the theoretical principles behind spectral information and are aware of the limitations of sensors and the sampling requirements for achieving acceptable levels of accuracy in classification. Likewise, knowing how to obtain reliable data from interviews in order to learn firsthand about the variability in human behavior and culture, as well as in plant distributions and stand structure, makes the remote sensing specialist more realistic about what can be brought back from the field and the reasons for trying to represent classes of local economic or environmental interest. It is common among remote sensing practitioners to perform field work largely to check the accuracy of categories that emerge from spectral analysis or to give them names, without much interest in changing classes that may prove to correspond poorly to field reality. The feedback from social science research and the incorporation of culturally meaningful categories are important contributions to the mapping of changes between land-cover classes and understanding of the driving forces of such changes. One of the important reasons for social scientists to play an increasing role in efforts to classify land uses with remotely sensed data is that in so doing they can begin to shift the applications of remote sensing from a mapping mode to one that seeks to explain social structures and processes—land-use dynamics, the lag time between commodity price shifts and landscape transformations, the estimation of yield from noncereal and even agroforestry crops, and the internal structure of households as revealed by the behavioral outcomes of that structure that are visible in land-cover changes.

Thus whenever possible, graduate students should be encouraged to pursue a global change graduate minor that will give them at least minimal combined exposure to the theories and methods of the social sciences, the biological sciences, and remote sensing. Lacking this, private foundations and federal agencies should be receptive to institutions wishing to support intensive training programs designed to introduce faculty and graduate students to these skills so they can participate effectively as members of multidisciplinary teams on global change. Indiana University, with support from the Tinker Foundation, has provided such training for the past 2 years to Brazilian and Mexican colleagues. Over the next several years, the NSF-funded Center for the Study of Institutions, Population and Environmental Change at Indiana University will conduct summer institutes addressing these needs, as well as continue to offer visitors monthlong individualized training linking social science and remote sensing to questions of environmental monitoring and global change, particularly in forested environments.

It may be hoped that other institutions will seek additional ways of meeting the challenge of addressing what sometimes becomes a major obstacle to collaboration between social scientists and those in the remote sensing community: a lack of familiarity with the techniques of the other field and with how their distinct but complementary skills can be linked to address questions of joint interest. The role of NSF in supporting multidisciplinary work through its Program on the Human Dimensions of Global Change has been crucial to advances made in this area to date. Continued support for such work by NSF and other agencies, such as the National Institute of Child Health and Human Development (NICHD), the National Aeronautics and Space Administration (NASA), and the National Oceanic and Atmospheric Administration (NOAA), would help in further eliminating existing barriers and creating a fertile ground for additional contributions to a basic understanding of human impacts on a changing environment.

In addition to a lack of familiarity with one another's skills, another notable obstacle to collaboration between social scientists and experts in remote sensing is the continuing high price tag for obtaining high-resolution satellite data, such as data from TM and the French Système Pour l'Observation de la Terre (SPOT). Despite regular promises to the community that the price will soon return to "the price of acquisition," many obstacles remain. While archived data have come down in price, more recent TM scenes still cost over $2,000, and the recent TM scenes have had serious problems involving uncertainties over sensor calibration that make such an expense increasingly risky. A spatial resolution of 30 m or better is necessary to address many of the human-dimension questions of concern to the social sciences, yet the discussion of Earth Observing System (EOS) instrumentation has seldom included input from social scientists to ensure that concerns related to the human dimensions of global change would be given high priority in sensor design. The remote sensing community needs to develop a mechanism for liaison with the social science community that engages them both in the decision-making process regarding the kinds of earth-observing instruments needed to understand human impacts on the environment at a variety of scales. Coarse scales tend to mask both environmental and human variability— one of the main things threatened by environmental change. To understand human and biological diversity, we need instrumentation that is sensitive to these fine-scale patterns and permits the linkage of fine-scale field research to remotely sensed data (see Cowen and Jensen in this volume, who raise similar issues relevant to work in urban areas). The other notable constraint is cloud cover and shadow, the latter especially in mountainous areas. Advances in radar technology should help with this problem since clouds do not interfere with radar. However, applications of radar data to land-cover analysis will require considerable technical advances before they can be used profitably by the social sciences.

A new generation of sensors is expected in the near future (e.g., Earlybird, Quickbird, the ASTER thermal emissions radiometer, Moderate Resolution Imaging Spectroradiometer [MODIS]). It is likely that social scientists can help evaluate the capabilities of proposed sensors, as well as contribute to the discussion of data availability. As important as it is to improve spatial resolution, there needs to be a commitment by remote sensing institutions responsible for data recording that information will be stored for regions all over the world. Images from SPOT sensors, for instance, despite their higher spatial resolution, have restricted availability for isolated areas. This results in a spotty land-cover change record that renders multitemporal analysis limited in scope.

In terms of spectral resolution, there is a need to discuss the possibility of dividing bands such as TM4 and TM5 into smaller spectral regions, and to evaluate whether and how such a change could improve future studies of land-use and land-cover changes, especially those related to agriculture. By the same token, it is still unknown what kind of information would be available if a thermal band (such as TM6) were designed at a higher spatial resolution, such as 30 or 20 m. Landsat 7, which will be launched in 1998, is expected to have a 30-m thermal band.

One point that is of particular importance to remote sensing analysis but is frequently dismissed is the need to work with digital data that have been calibrated (i.e., converted to reflectance values). The implementation of technical procedures for performing such calibrations requires considerable expertise found only rarely outside of major remote sensing facilities. There would appear to be enough capability within remote sensing institutions responsible for data reception to develop relatively automated procedures that would facilitate this kind of data preparation for those, like social and biological scientists, who lack this level of laboratory or technical expertise. Even some in the remote sensing community use other procedures to get around the complex uncertainties involved in transforming digital numbers to reflectance values.

In terms of software development, there should be continuous support for the development of low-cost, interactive, yet powerful packages, such as IDRISI (developed by Clark University) and MULTISPEC (developed by Purdue University). Such packages should include a range of statistical tools allowing analysis of data for training samples to be used in supervised classification and determination of the accuracy of thematic maps. Such software packages provide an ideal tool for training social scientists, since they allow more effort to be dedicated to image analysis and interpretation than to the learning process for the software itself. Many social scientists are reluctant to work with digital data because of the slow learning curve for many remote sensing and geographic information system (GIS) packages, which translates into virtual inaccessibility of image processing to its numerous potential users. The growing capability of personal computer (PC) processors now frees new members of the community from the need to rely on the UNIX platform for working with these data. Even the powerful ERDAS Imagine image processing software is now available for the PC at a cost that is affordable under the most modest of grants.

CONCLUSIONS

Looking back over the past 5 years of our project linking detailed field studies to remotely sensed data, we believe that on the whole, this joint work has advanced our knowledge of important processes of land-cover change in Amazonia and our knowledge of how to link data across levels of analysis, and holds promise for further contributions in the immediate future. Without our detailed field inventories and land-use histories, we would not have been able to distinguish among three distinct stages of secondary succession in the TM images (Mausel et al., 1993; Moran et al., 1994), to distinguish floodplain forest from açaí palm managed floodplain forest (Brondizio et al., 1994, 1996), to demonstrate the linkage between soil fertility and rates of secondary succession at a more than highly localized scale (Moran et al., 1996), and to determine the relative impacts of various land-use trajectories in specific soils on species composition and biomass accumulation (Moran et al., 1996; Tucker et al., in press). The use of satellite remote sensing modified our approach to sampling so that it became more widely distributed over the landscape than it would otherwise have been. This brought us into contact with households, soils, and landscape patterns different from those we would have sampled if we had relied on traditional techniques. In turn, these contacts enhanced the regional scope of our conclusions about land use and its impact on land cover and produced statistics for areas much larger than would otherwise have been possible, while our detailed field studies allowed us to modify land-cover classes and provided enhanced discrimination. Cumulatively, our various studies have developed structural criteria that facilitate the application of these considerations in spectral analysis of satellite data for other Amazonian regions.

The linking of remotely sensed data to traditional field methods in the social and biological sciences has permitted more thoughtful sampling over a larger region, addressing questions of decadal change that could not be examined through traditional methods alone. We have documented land-cover change at five separate Amazonian locations in 5 years—a task that 10 years ago we would have thought impossible even to consider. This change has been tracked with detailed quantitative measures and accuracies of 85 to 94 percent that provide levels of confidence rarely achieved by the traditional methods of the social sciences.

Our work in Marajó, for instance, has changed how we conceptualize Caboclo populations and their engagement in the regional economy. Study of the intensification of flood plain agroforestry through a combination of household interviews, field inventory, and image classification of these areas has shown that the areal extent of agroforestry stands represents the most important production system in the region. On the one hand, it changes the characterization of the population from extractivists to intensive farmers—forest farmers. Furthermore, it shows that the level of food production can be increased without increasing rates of deforestation (Brondizio, 1996; Brondizio and Siqueira, 1997).

One can also examine changes in the densely populated Bragantina region. This area, characterized for 100 years as an area dominated by smallholders, commonly on 25-ha farms, has been undergoing transformation in recent years. What is the extent of this transformation? Is it taking place only near towns or along major roads, or is it pervasive? The use of remote sensing can help monitor these changes in land cover and help us question their desirability in social and environmental terms. At a time when cities such as Belém more than ever need a green belt to supply them with produce, the traditional sector that has supplied it may be disappearing as a result of uninformed policies or lack of support.

Altamira likewise is a landscape that poses many questions for social and environmental scientists. Will it begin to experience the same kind of logging-related fires that have plagued the Paragominas region and Borneo? Our recent work on the structure of households in the area using a property-level grid over-laid with satellite image data suggests a rapid expansion of logging beginning in 1985 and accelerating in 1988. Whereas logging was concentrated within the first 25 km from the town, in 1985, by 1991 loggers were altering areas more than 75 km from the town and desirable species have become increasingly rare closer to town. What is the spatial distribution of pasture and other economically significant land uses? What is the impact of road distance and road quality on economic activity? In some preliminary work with the near- and mid-infrared bands of TM, we detected a distinct pattern of land use in which the higher-ground farms experienced greater deforestation and land use than those occupying lower positions in the landscape. Is this a product of soil type differences along a soil catena, or of moisture saturation in lower sites? Can analyses using infrared bands provide a quick way of identifying better soils for agriculture in newly settled areas?

In the Japanese colony of Tomé-Açú there is evidence of incipient change toward the expansion of cattle ranching and away from the intensive systems of production that have characterized the past 65 years of occupation. The combined tools of socioeconomic analysis and remote sensing can provide a means of effectively monitoring such changes in culture and society that are of considerable theoretical and practical significance in terms of regional development. Tomé-Açú has long been seen as offering an example of an alternative to cattle ranching in Amazonia. Understanding of how and why this human community is shifting to cattle has considerable economic and environmental significance.

Perhaps more important to discussions of global environmental change are findings that show the large extent of carbon sequestration by secondary vegetation and its very high rate in the initial 10 years after abandonment, its spatial variability as a function of soil fertility, and the role of land use in this process (Randolph et al., 1996). In documenting the role of soils we have also found some counterintuitive results, such as very large carbon pools in the soil and in both surface and deep roots. Contrary to past wisdom suggesting that the root systems of tropical moist forests are shallow, we see the legacy of roots deep in the soil profile (Nepstad et al., 1994).

The results of the studies discussed here have led us in some new research directions, including examination of the role of the developmental cycle of domestic groups in shaping the trajectory of land use and deforestation in frontier regions and the role of community-level organizations in managing forest resources. Household composition may explain differential rates of deforestation through time better than current models focusing on migrant origins and flows. The impact of age and gender composition on strategies of land use is being examined with support from NICHD. This research will elucidate the impact of aggregate migration flows relative to that of household types within the migrant pool and the changing behavior of households through time as they mature and change in composition. A second line of investigation, under the NSF-funded Center for the Study of Institutions, Population and Global Change at Indiana University, is incorporating the community level of organization into our studies—a level that falls between the household and landscape levels on which we have focused in the past. At this level, we will be examining the organization of user groups within communities and the observable differences among groups using forest resources within different property regimes and demographic spatial distributions. This level will be linked to the field-level and landscape-level data discussed in this chapter.

In both of the above new efforts, remotely sensed data play a key role at every stage of the research—from the exploration of types of land cover, to the sampling approach taken in the field work, to the interviews with land users, to the analysis of land-use changes in time and space. These plans suggest a productive collaboration among social scientists, biological scientists, and the remote sensing community for years to come.

ACKNOWLEDGMENTS

The authors thank the National Science Foundation, which through grants 91-00526 and 93-10049 has provided support for these five Amazonian land-use studies. Support for carbon modeling has been provided by the Midwestern Regional Center of the National Institute for Global Environmental Change. Support to the second author from the Indiana Center for Global Change and World Peace and NASA's Global Change Fellowship Program (3708-GC94-0096) is gratefully appreciated. The work reported on here is the product of the efforts of many on our team. From the United States are J. C. Randolph, Amy Gras, Alissa Packer, JoAnne Michael, Joanna Tucker, M. Clara da Silva-Forsberg, Fabio de Castro, Stephen McCracken, Warren Wilson, Cindy Sorrensen, Bernadette Slusher, Vonnie Peischl, and Masaki Yamada. From Brazil are Mario

Dantas, Italo Claudio Falesi, Adilson Serrao, Lucival Rodrigues Marinho, Jair da Costa Freitas, Therezinha Bastos, and many others in collaborating institutions— especially Empresa Brasileira de Pesquisa Agropecuária/Centro de Pesquisa Agroflorestal do Trópico Únido (EMBRAPA/CPATU). We extend our appreciation also to Ronald Rindfuss, who provided valuable comments, and to Paul Mausel and the Department of Geography and Geology at Indiana State University, especially the staff of their Remote Sensing and GIS Laboratory, for wise counsel at many stages of our work. We also wish to thank the many local people who answered our questions about their uses of land with such good humor and insight. The views expressed herein are the sole responsibility of the authors and may not reflect the views of our funding sources or our collaborators.

NOTES

1 A notable exception is efforts by Woodwell et al. (1986, 1987) to differentiate between mature tropical forest and secondary succession. That early effort, unfortunately, encountered the limitations posed by the spatial resolution of the Landsat Multispectral Scanner.

2 Among those who have pursued the acquisition of these skills are Clifford Behrens, Bruce Winterhalder, the late Robert McC. Netting, Endre Nyerges, and Emilio Moran. There may be others that have escaped our notice. Others have preferred conducting cooperative work with remote sensing specialists (e.g., Conrad Kottak) or taking advantage of courses offered at their universities (e.g., George Morren and Thomas Rudel). The choice of whether to seek such training oneself or rely entirely on the expertise of collaborators is an important one that reflects personal style, role on the team, and synthesis goals. This is a mechanism that, if used by other social science disciplines, could substantially increase the number of scholars engaged in this type of work—although there are other modalities for achieving this goal that are discussed later in the chapter.

3 Our work would not have been possible without collaboration with the remote sensing group at Indiana State University's Remote Sensing and Geographic Information Systems Laboratory. Collaboration with Paul Mausel has been particularly valuable over several years.

4 The importance value is a measure of relative dominance, frequency, and density.

5 GPS devices provide accurate location through triangulation using at least 3, and often more, of the 24 satellites in the system.

6 Color composites made up of Landsat TM bands 5 (mid-infrared), 4 (near-infrared), and 3 (visible) provide a very realistic picture of the landscape that facilitates their field use in interviews.

7 During field work we use a "synthesis table" containing the structural characteristics of a variety of land-cover classes to guide our training sample data collection. This synthesis table includes information such as average stand height and range of diameter at breast height for particular land-cover classes to facilitate discrimination of classes of interest. This guide is based on earlier field vegetation inventories carried out in the region, but it can also be based on existing studies carried out by others.

REFERENCES

Alvim, P. de T.
1974   Um Novo Sistema de representação gráfica da fertilidade de solos para cacau. *Cacau Atualidades* 11(1):2-6.
Behrens, C., M. Baksch, and M. Mothes
1994   A regional analysis of Bari land-use intensification and its impact on landscape heterogeneity. *Human Ecology* 22(3):279-316.
Booth, W.
1989   Monitoring the fate of forests from space. *Science* 243:1428-9.
Brondizio, E.
1996   Forest Farmers: Human and Landscape Ecology of Caboclo Populations in the Amazon Estuary. Ph.D. dissertation, Indiana University, Bloomington.
Brondizio, E., E. Moran, P. Mausel, and Y. Wu
1994   Land-use change in the Amazon estuary: Patterns of Caboclo settlement and landscape management. *Human Ecology* 22 (3):249-278.
1996   Land cover in the Amazon estuary: Linking of the thematic mapper with botanical and historical data. *Photogrammetric Engineering and Remote Sensing* 62 (8):921-929.
Brondizio, E., and A. Siqueira
1997   From extractivists to forest farmers: Changing concepts of Caboclo agroforestry in the Amazon estuary. *Research in Economic Anthropology* 18:233-279.
Buschbacher, R.J., C. Uhl, and E.A.S. Serrao
1988   Abandoned pastures in eastern Amazônia (Brazil): II. Nutrient stocks in the soil and vegetation. *Journal of Ecology* 76:682-699.
Denich, M.
1991   Estudo da importancia de uma vegetação nova para o incremento da produtividade do sistema de produção na Amazônia Oriental. Ph.D. dissertation, EMBRAPA/GTZ, Belém, Brazil.
Lathwell, D.J., and T. L. Grove
1986   Soil-plant relationships in the tropics. *Annual Review of Ecology and Systematics* 17:1-16.
Lean J., and D. Warrilow
1989   Simulation of the regional climate impact of Amazonian deforestation. *Nature* 342:411-412.
Li, Y., P. Mausel, Y. Wu, E. Moran, and E. Brondizio
1994   Discrimination between advanced secondary succession and mature forest near Altamira, Brazil, using Landsat TM data. Vol. 1, pp. 350-364 in ASPRS/ACSM Annual Convention and Exposition ASPRS Technical Review Papers. Bethesda, Md.: American Society for Photogrammetry and Remote Sensing and America Congress on Surveying and Mapping.
Mausel, P., Y. Wu, Y. Li, E. Moran, and E. Brondizio
1993   Spectral identification of successional stages following deforestation in the Amazon. *Geocarto International* 8 (4): 61-71
Moran, E.
1976   *Agricultural Development along the Transamazon Highway.* Center for Latin American Studies, Monograph Series #1, Indiana University.
1981   *Developing the Amazon.* Bloomington: Indiana University Press.
1993   *Through Amazonian Eyes: The Human Ecology of Amazonian Populations.* Iowa City, Ia. University of Iowa Press.

Moran, E., E. Brondizio, P. Mausel, and Y. Wu
   1994   Integrating Amazonian vegetation, land use and satellite data. *BioScience* 44 (5): 329-338.
Moran, E., A. Packer, E. Brondizio, and J. Tucker
   1996   Restoration of vegetation cover in the eastern Amazon. *Ecological Economics* 18:41-54.
Murrieta, R., E. Brondizio, A. Siqueira, and E. Moran
   1989   Estrategias de subsistencia de uma população ribeirinha da ilha de Marajó. *Boletim do Museu Paraense Emilio Goeldi* (N.S. Antropologia) 5:147-163.
Nepstadt, D., C.R. de Carvalho, E. Davidson, P. Jipp, and P. Lefebvre
   1994   The role of deep roots in the hydrological and carbon cycles of Amazonian forests and pastures. *Nature* 372 (6507):666.
Randolph, J.C., R.B.Slusher, E. Moran, and E. Brondizio
   1996   Primary production, litterfall, and decomposition in second growth forest in the Eastern Amazon. Paper presented at the ESA annual meeting, Providence, R.I. *Bulletin of the Ecological Society of America* 77:367.
Skole, D., and C.J. Tucker
   1993   Tropical deforestation and habitat fragmentation in the Amazon: Satellite data from 1978 to 1988. *Science* 260:1905-1910.
Smith, N.
   1982   *Rainforest Corridors.* Berkeley: University of California Press.
Sombroek, W.
   1984   *Amazon Soils.* Wageningen, The Netherlands: Bureau of Agricultural Research.
Tucker, J.M.
   1996   Secondary Succession in the Brazilian Amazon: Investigation of Regional Differences in Forest Structure in Altamira and Bragantina, Pará State. Honors thesis, Indiana University.
Tucker, J.M., E. Brondizio, and E. Moran
   in   Regional Differences in the Rate of Secondary Forest Succession: A Comparison of
   press   Altamira and Bragantina. *Interciencia*
Vieira, I., D. Nepstadt, R. Salomão, J. Roma, and N. Rosa
   1996   Regio Bragantina: as florestas secundaries apos um seculo de agricultura na Amazônia. *Ciencia Hoje* 20(119):38-44.
Woodwell, G., R. Houghton, T. Stone, and A. Park
   1986   Changes in the areas of forests in Rondonia, Amazon Basin, measured by satellite imagery. Pp. 242-257 in *The Changing Carbon Cycle: A Global Analysis*, J.R. Trabalka and D.E. Reichle, eds. New York: Springer-Verlag.
Woodwell, G., R. Houghton, T. Stone, R. Nelson, and W. Kovalick
   1987   Deforestation in the tropics. New measurements in the Amazon Basin using Landsat and NOAA advanced very high resolution radiometer imagery. *Journal of Geophysical Research* 92:2157-2163.

6

Land-Use/Land-Cover and Population Dynamics, Nang Rong, Thailand

Barbara Entwisle, Stephen J. Walsh,
Ronald R. Rindfuss, and Aphichat Chamratrithirong

This chapter describes ongoing research in Nang Rong, Thailand that joins social, biophysical, and spatial perspectives, data, and tools to study population and environment there. To date, this research has addressed two sets of questions:

• Did land use/land cover in the 1970s and early 1980s affect the subsequent out-migration of young adults (Rindfuss et al., 1996a)? More specifically, did the availability of relatively undeveloped forested land close to villages act as a brake on out-migration that otherwise would have occurred? Did fragmentation of land use and potential competition from residents of other nearby villages have an effect?

• Did population change between 1984 and 1994—including rates of growth (or decline), household formation, and net in- and out-migration—affect land use/land cover in 1994 (Walsh et al., in press; Entwisle et al., 1997b)? If so, what was the nature of these effects?

The purpose of this chapter is to use our research experience to consider the challenges of the multidisciplinary and integrative approach we have taken. Accordingly, we discuss the history of our projects, the intensive focus on a relatively small geographic area, the development of a prospective multilevel survey-based social data set, the creation of a complementary satellite time series, the choice of villages as the primary link between data sources, and other issues of scale compatibility.

NANG RONG, THAILAND: LABORATORY FOR STUDY

Nang Rong district, Thailand occupies approximately 1300 km^2 in the southern part of the Korat Plateau in Buriram Province in the Northeast of the country (see Figure 6-1). The idea that this relatively small geographical area (approximately the size of a U.S. county) should be the focus for intensive study over a broad range of interrelated topics is key to our work. The "laboratory" approach has been used quite successfully by others interested in social change (for example, the Middletown studies [e.g., Caplow et al., 1982]) or demographic change (for example, the Matlab project [Phillips et al., 1988]), although it is the exception rather than the rule in social demographic research. Our goal is a comprehensive account of social, economic, demographic, and environmental change within Nang Rong district. Our approach is cumulative. One element at a time, we have built a database integrating diverse sources (social surveys, administrative data, map products, remote imagery) that is substantively broad, includes multiple levels of observation, and is unusual in its temporal depth. This database is continually expanding as new sources are identified and new data acquired. In parallel fashion, the knowledge base is also in a state of continual expansion, as each study builds on preceding ones. We believe the laboratory approach has much to offer for the study of population and environment.

Why was Nang Rong selected? The origins of our projects lie in a development project and in data collection to evaluate its impact. In 1984, the Population and Community Development Association (PDA) began a Community-Based Integrated Rural Development (CBIRD) project in selected villages in the district. The CBIRD project was designed to (1) improve skills and productive capacity in agriculture, animal husbandry, and cottage industry, and (2) upgrade waste disposal facilities, increase year-round availability of clear water, and promote improved individual health care practices. Nang Rong was selected because it was (and is) a relatively poor district in an historically poor region of Thailand. "Nang Rong" means "sit and cry."

The CBIRD project and associated data collection provided the base for our projects, but what sustained our interest was the multifaceted nature of change under way in Nang Rong district. The past few decades have witnessed a major fertility decline, increased use of contraception, electrification, the extension and improvement of the road network, increasing frequency of bus service, more widespread use of tractors (mostly "walking tractors") for land preparation, an increase in the number of mechanized rice mills, and improvements in sanitation and water storage. Initially, we were interested in contraceptive choice within a changing environment; we later expanded our scope to consider migration and fertility in the context of change. With the broadening of our substantive interests, along with additional data collection (described below), the work that began as an evaluation of the CBIRD project grew until Nang Rong became a laboratory for the study of social and demographic change.

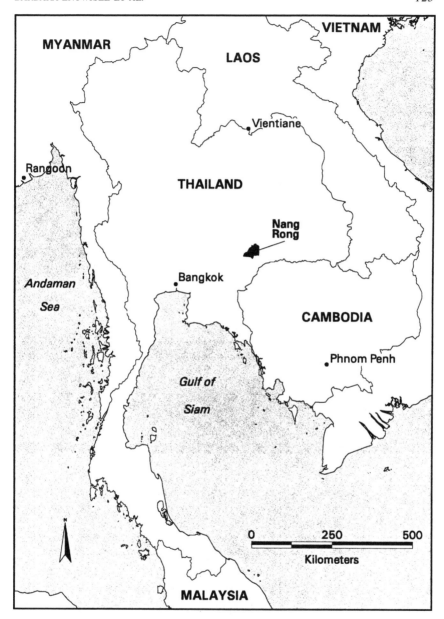

FIGURE 6-1   Location of Nang Rong study area.

Given the concept of Nang Rong as a laboratory, extending our work to include environmental concerns was a natural step. The area that makes up the district is noted for its undulating landscape; in the lower parts there is paddy land, and in the higher parts open-canopy forest and upland for dry field crops (Fukui, 1993). As with the Northeast as a whole, the region is composed of Cretaceous sandstone, shale, and siltstone overlaid with Tertiary and Quaternary alluvial deposits that have been eroded to form a succession of low-lying terraces (Pendleton, 1963; Donner, 1978). As is typical of the Korat Plateau, the soils are dominated by fine sandy loams that form one of the most infertile soil groups in Thailand. Cultivation of wet rice occurs in shallow depressions, low terraces, and alluvial plains, while dry dipterocarp savanna forests or drought-resistant crops are grown on the upland sites (Parnwell, 1988). Annual rainfall is low, leading to low rice yields and interannual instability (Limpinuntana et al., 1982; Fukui, 1993). Precipitation is also variable. While nearly 80 percent of the annual total falls within a 5-month period from May to September, the rains may arrive in early March or well into June or July. September may be a dry month, or it may be so wet that extensive flooding occurs. Droughts are not uncommon in the middle of the rice-growing season, June to September. Precipitation extremes are the rule rather than the exception in Nang Rong.

Major changes in land use have occurred over the past several decades, paralleling social, economic, and demographic change. Agricultural productivity grew rapidly between 1960 and 1980, mostly because of the rapid extensification of agricultural land into formerly forested and upland areas (Siamwalla et al., 1990). In the Northeast as a whole, 60 percent of the land was forested in 1960, but only 15 percent in 1982 (Panayotou and Sungsuwan, 1989). Cropping patterns also changed. Marginal lands were cleared initially to grow cassava, exported to Europe as a calorie rich ingredient for livestock feed (Phantumvanit and Sathirathai, 1988). In recent years, extensification has slowed as a result of preventive deforestation policies, demands for surplus labor in the rapidly growing manufacturing and urban sectors (Tambunlertchai, 1990), and the availability of only marginal lands for further agricultural expansion. The interaction of population and environment through forces within and exogenous to the region has created a dynamic landscape mosaic.

Nang Rong thus offers an important context in which to study interrelationships between population and environment, and the already existing data and knowledge base helped justify the considerable additional investment required to establish a geographic information system (GIS), acquire and process remote imagery, and integrate social survey and remotely sensed data in analyses of population change and land use/land cover.

PROJECT HISTORY

In the fall of 1992, we collaborated on our first grant proposal, submitted to the National Science Foundation (NSF), to integrate spatial data, methods, and perspectives into ongoing social and demographic research in Nang Rong. Sara R. Curran, now of Princeton University, contributed importantly to this effort. With NSF funds, we accomplished several important goals. First, we developed the basic elements of a GIS to link spatial and social survey data. Second, we purchased, processed, and evaluated five frames of Landsat satellite data from the 1970s. Details are given below, but for purposes of describing the project history, suffice it to say that the satellite data provided an initial look at land use/land cover in the district, showing, for example, the dominance of rice cultivation, spatial variability in crop mix and landscape pattern, and change over time. The promise of Nang Rong as a site for research on population and environment was thus borne out in the initial remotely sensed data. Third, and more substantively, we initiated a study of out-migration in relation to land use/land cover (Rindfuss et al., 1996a).

While the NSF proposal was under review, we developed a related initiative to integrate social and spatial data in the evaluation of family planning programs. Funded as a subproject under the EVALUATION Project, sponsored by the U.S. Agency for International Development, with additional funds from the Mellon Foundation, this effort, led by Amy Tsui, explored the GIS as a means of organizing and integrating family planning program data from multiple sources and time points. Alternative measures of family planning accessibility were derived from spatial data using spatial network analysis, and then incorporated into statistical analyses of contraceptive choice that drew on measures based on social survey and administrative data as well (Entwisle et al., 1997a). Although not specifically focused on population and environment, these analyses helped further the GIS and our collaborative strength in the marriage of social and spatial data, techniques, and perspectives.

Consequently, when the National Institute of Child Health and Human Development (NICHD) issued its Request for Applications on "Population and Environment" in June 1994, we were well positioned to compete for funds. We had already created a data set for Nang Rong that was unique in its social, spatial, and temporal coverage, and a program of social demographic research based on these data was well under way (Entwisle et al., 1996; Rindfuss et al., 1996b). Instrumental to this latter research was an initial grant received from the National Institutes of Health (NIH) to study contraceptive choice in a changing environment based on 1984 and 1988 data, which was followed by a competing continuation grant to collect additional social data in 1994-1995, investigate social networks in relation to other characteristics of social context, and expand the scope to include migration as well as fertility. A GIS was in place. We had processed satellite data and could demonstrate their relevance to a study of land-use/land-cover and population change. An interdisciplinary team was in place that had collaborated successfully and developed some common language.

Our project, "Population Dynamics and Changes in the Landscape," was funded and is under way at the time of this writing. The first goal was to expand the time series of optical satellite images for Nang Rong from 1972 through 1996, and also to incorporate radar data and aerial photography extending back to the 1950s. The other goals of the project revolve around use of the satellite time series, combined with data from social surveys, administrative records, and digital maps, to explore dynamic interrelationships among land use, population, and social and economic change. As indicated earlier, these explorations have focused thus far on land use/land cover in the 1970s and early 1980s as an influence on subsequent out-migration of young adults, and on population change between 1984 and 1994 as an influence on land use/land cover in 1994. Next steps include the development of a household model linking land use and migration, a study of the consequences of past high fertility for the current size of landholdings and for the fragmentation of agricultural plots, and the construction of models for assessing landscape organization and use of these models to simulate trajectories of future change.

A related project, funded by the MacArthur Foundation, has focused specifically on villages as the link between measures derived from the satellite time series and those derived from social survey and administrative data. The maps available to that project are based on different concepts of "village," and we are just now completing the field work necessary to resolve this problem. The 1984 topographic map identifies clusters of dwellings as villages, whereas the 1993 planning map identifies administrative units. This difference in concept, combined with real change over the decade separating the maps, led to substantial discrepancy between the two maps. Each map shows locations for approximately 300 villages, but only 200 of those villages appear in both maps. Accounting for and resolving these differences is vital for analyses of land use/land cover and population over time.

We have described the history of our Nang Rong projects in terms of each project building on previous ones, but this is not the whole story. An expanding research infrastructure for spatial analysis paralleled the progression of individual research projects. There is a division of labor between the Carolina Population Center (CPC) and the Spatial Analysis Laboratory in the Department of Geography at the University of North Carolina with respect to data processing, but the nature of this division of labor has changed over time. Initially, social data processing took place at CPC, and spatial data processing took place at the Spatial Analysis Laboratory. Gradually, CPC's Spatial Analysis Unit was developed, first with start-up funding from the university's Vice Chancellor for Health Affairs and then with funding from an NICHD "Center" grant. We now use both spatial analysis units, which has increased efficiency and allowed us to draw on a wider range of expertise.

DATA

Social Survey Data

Key to our research on land-use/land-cover and population change is prospective longitudinal and multilevel survey data, with observations for individuals, households, and villages covering the 1984-1994 period. As mentioned, the longitudinal data collection began as an evaluation of the CBIRD Project (Institute for Population and Social Research, 1984:3). To evaluate the success of the initial 4-year project, the Institute for Population and Social Research (IPSR), Mahidol University, Thailand, with financial assistance from the Canadian International Development Research Center, conducted a complete household census in each of 51 villages in 1984, a 1988 follow-up survey of households containing a woman of reproductive age, and community surveys in the 51 villages at both time points as part of a comprehensive research effort. These are known as the CEP (Community Evaluation Project) surveys. No U.S. researchers were involved in the 1984 surveys, but Ronald R. Rindfuss and David K. Guilkey of CPC collaborated with Aphichat Chamratrithirong and other IPSR researchers to design the 1988 data collection. The idea was to piggyback a study of contraceptive choice onto the CBIRD evaluation already planned (Entwisle et al., 1996; Rindfuss et al., 1996b). The linked data sets have also been used to study migration (Curran, 1994; Sawangdee, 1995).

Likewise, the 1994 surveys (funded by an earlier NIH grant) built on the foundation of the 1988 survey panel. These surveys, known as the CEP-CPC surveys, followed up individuals, households, and villages included in the 1984 data collection, but with a greatly expanded research agenda. The 1994 data were intended to support prospective migration studies, detailed descriptions of life-course transitions in the young adult years, and studies of social networks at multiple levels of observation (individual, household, village) in relation to demographic behavior. Four different theoretical perspectives guided this effort: Davis' (1963) multiphasic theory, which stresses multiple demographic responses to social change (especially the interconnectedness of fertility and migration); multilevel approaches to examining social contexts and demographic behavior (Entwisle and Mason, 1985; Bilsborrow et al., 1987; Findley, 1987); the life-course perspective, with its interest in individual role trajectories over time (Clausen, 1972; Elder, 1985, 1991); and social network theory (e.g., Freeman, 1979)—as applied to demographic phenomena (Rogers, 1983; Massey and Garcia Espana, 1987). There are three components to the 1994 data collection: a community profile, a household survey, and a migrant follow-up.

The 1994 community profiles were done in all of the Nang Rong villages, including but not limited to ones that were initially part of the CBIRD evaluation. The community profiles done in 1984 and 1988 were fairly extensive. The 1994 data collection obtained much of the same detail, including population size and composition; cropping (e.g., specific crops grown, prices received, marketing); water sources (location, quality, and sufficiency); agricultural technology (use of various kinds of equipment, fertilizers, and pesticides); electrification; transportation and communication; health and family planning services; and village groups and committees.   In addition, the 1994 community profile covered some new topics, including links to other villages by virtue of shared temples, schools, and bus routes; labor exchanges between villages; and perceptions about deforestation.

Although not specifically designed to be used with remotely sensed data, the village profiles provide data useful for many purposes. First, survey information about cropping, use of fertilizer, water sources, and deforestation provides a cross-check on the interpretation of remotely sensed data (and vice versa). Second, information about links and exchanges between villages can shed light on questions of scale (see below). Although the village seems to be an appropriate level at which to conduct both social and spatial analysis, it is possible that some clustering of villages may be better. Social network analysis can be used to develop a basis for such aggregation. Third, village profile data can be linked with variables derived from remotely sensed data, and statistical analyses can then be undertaken. We have conducted some preliminary analyses of population, land use/land cover, and landscape diversity along these lines (Walsh et al., in press; Entwisle et al., 1997b).

A household survey was the second component of the 1994 data collection. In fact, the survey consisted of a complete household census in each of the 51 villages that participated in the 1984 and 1988 data collections. The 1994 survey addressed the topics covered previously, and considerably more: substantial detail on the whereabouts and current characteristics of 1984 household members; visits and exchanges of goods and money with former household members; social and demographic facts about current members; yearly life history data for those aged 18-35, including information about migration patterns and occupation; sibling networks for those aged 18-35; and household characteristics, including plots of land owned and rented, use of agricultural equipment, crop mix, planting and harvesting of specific crops (rice, cassava, and sugarcane) and household debts.

These data have quite a bit to offer when used in conjunction with remotely sensed data. Aggregated to the village level, household data offer an additional perspective on the interpretation of satellite images. Data on crop mix and on the timing of planting and harvesting can be used to verify land-use/land-cover classifications based on satellite data for the same time period. It is also possible to estimate the total amount of land cultivated by village households, and this information can be used to adjust spatial boundaries around villages. As explained below, one of the challenges faced in joining social survey and remotely sensed data is the identification of a common unit of observation. Thus far, the village has served as our common unit. This approach has worked reasonably well, but a difficulty is that the spatial boundaries of a village are not so easy to determine.

Tom Evans of the University of North Carolina Department of Geography is working on a dissertation that will help clarify questions about boundaries in the Nang Rong setting. In the meantime, however, clues provided by the aggregated household data as to which villages are quite large spatially and which are quite small add realism to some of the arbitrary procedures we must employ.

Another key strength of the household data is the prospective design. Using the 1984 household rosters as a basis, survey interviewers in 1994 attempted to find out about all households and household members present in 1984. Identifying information was obtained so that the 1994 and the 1984 records could be linked. This means that for the 51 villages in which the household censuses were done, detailed data are available not only about population change over the 10-year period, but also about changes in population composition (for example, age, education, occupation, assets). Further, it is possible to differentiate between changes to households and household members present in 1984 and changes due to new individuals moving into the village and to the formation of new households. Most important, it is possible to examine population processes prospectively. We have conducted a preliminary analysis in which the out-migration of young adults between 1984 and 1994 is examined in relation to the availability of undeveloped land (based on remote imagery from the mid-1970s), the fragmentation of land use (also based on remote imagery), and the potential for competition from other villages (Rindfuss et al., 1996a). During this period, marginal lands were cleared to grow cassava in Nang Rong (and elsewhere in the Northeast) as part of the rapid extensification of agricultural land into formerly forested and upland areas. Consistent with this picture, we found that the availability of forested land in the 1970s was associated with diminished out-migration in the 51 villages where household data were collected. The prospective design has also made it possible for us to consider household formation, as well as changes in household size and composition, as a potential influence on land use/land cover (Entwisle et al., 1997b).

The final component of the 1994 survey is a follow-up of out-migrants from 22 of the 51 villages. These 22 villages were selected randomly within strata created by cross-classifying general location (quadrant) and distance from major paved roads in 1984. Persons resident in 1984 but no longer resident in 1994 were candidates for follow-up if they had gone to Bangkok and surrounding areas, the Eastern Seaboard (a focus of rapid growth and development), Korat (a regional city), or Buriram (the provincial city). We succeeded in finding and interviewing about 70 percent of the migrants who still had family in Nang Rong, a remarkably high proportion (cf. Bilsborrow et al., 1984). In Nang Rong, migration of young adults is extremely common. Some return, while some do not. An understanding of migration patterns thus requires an understanding of the factors influencing some temporary out-migrants to become permanent out-migrants. Such an understanding in turn requires prospective and retrospective data on those who leave, as well as those who stay or have returned. The Nang Rong data are unique in providing such a complete picture of migration.

As we have stressed, the 1994 surveys were not designed specifically for use with satellite data[1] (and we would have done some things differently had we had the wisdom of hindsight), but collectively, and combined with the 1984 and 1988 data collections, the survey data offer excellent opportunities for analysis in conjunction with remotely sensed data. There are many features of the survey data that make their integration with satellite imagery especially exciting. One is coverage. As noted earlier, the 1994 community profiles were done in all of the villages, while data were collected from all households in a subset of 51 villages. A second important feature is the multilevel structure of the data: there are observations for individuals, households, and villages. With this structure, even though we cannot link individual people to individual pixels, it is still possible to study individuals and households in contexts characterized in part by satellite data. This is important because many land-use decisions are made by individuals and households. A third important feature is temporal depth. Ten years is not such a long time for interrelationships between population and environment to manifest themselves, but it is a longer period than that spanned by many social demographic data sets, especially prospective ones. A final and related feature is that the data follow migrants as well.

Remotely Sensed Data

Remote images obtained from the Landsat Multispectral Scanner (MSS) and Thematic Mapper (TM) and from the Système pour l'Observation de la Terre (SPOT) Multispectral (MX) and Panchromatic (PAN) sensors are also key to our research on land-use/land-cover and population change in Nang Rong. The following discussion briefly describes the approaches we are following to assess land use/land cover and plant biomass through satellite spectral responses. GIS techniques used to derive biophysical and geographical variables are also de-scribed, as are pattern metrics for assessing the composition and spatial organiza-tion of land use/land cover derived from the digital classifications.

Using a derived local crop calendar, we constructed a satellite time series that incorporates Landsat MSS, TM, and SPOT MX and PAN data. The MSS data include scenes for 18 December 1972, 28 February 1973, 24 November 1975, 17 January 1976, and 7 October 1979; the TM data include scenes for 17 November 1988, 6 February 1989, 15 December 1992, 2 February 1993, 3 No-vember 1994, 6 January 1995, 7 February 1995, 6 November 1995, 24 December 1995, and 25 January 1996; and the SPOT data include MX and PAN scenes for 7 April 1994. The time series was configured to provide (1) historical coverage prior to the 1984 population survey and subsequent to the 1994 survey; (2) multispectral, multispatial, and multitemporal representation; and (3) interannual and interseasonal characteristics for defining the nature of land use/land cover and the magnitude, pattern, and directionality of change. Additional Landsat TM and SPOT MX scenes will be acquired as our research proceeds.

Based on our experiences with processing of images for the region, the importance of a satellite time series for land-use/land-cover mapping has become clear. Variation in local and regional phenologies, planting schedules across the district, and land-use/land-cover types suggests the need for a minimum of two and preferable three images per analysis period for satisfactory land-use/land-cover separation. Following data preprocessing to correct for geometric, radiometric, and topographic distortions in the data, we used empirical normalization procedures to balance and correct responses from data acquired on different dates and therefore under different solar angles and atmospheric attenuation characteristics. The image time series was then classified for land-use/land-cover extraction. An unsupervised classification approach was used to map land use/land cover at multiple dates, with emphasis on 1979 and 1993. Level 1 and Level 2 mapping classes were defined through the classification process, with emphasis on the pattern of deforestation, agricultural intensification in rice-growing areas, and agricultural extensification of cassava in upland sites. Plate 6-1 (after page 150) provides an example. Supervised classification is also being used, as well as hybrid approaches, as our ground control information expands areally and across land-use/land-cover types, and as we seek to identify specific cover types for selected villages, landscape strata, or time periods.

The accuracy of a classification is summarized through an error matrix, Kappa statistic, and omission and commission errors by comparing satellite-based land-use/land-cover categories against aerial photography and/or field samples. We are using two approaches. The household and village surveys contain information on related variables, such as the total area planted in various crops, which can be compared with the Level 1 and Level 2 classes. Also, as a consequence of prior trips to the district, a ground-control GIS directory has been created and expanded over time. This directory contains assorted ground-control coverages that relate to various land-use/land-cover verification efforts conducted for various villages within the district. Locations are defined through Global Positioning System (GPS) values and through delineated plot locations on scaled Landsat TM and SPOT PAN satellite output.

Our initial work with these land-cover classifications has involved examining the relationship between migration patterns and patterns of land-cover change (Rindfuss et al., 1996a; Entwisle et al., 1997b), as well as more general relationships between population and land-cover patterns (Walsh et al., in press). As part of this work, we are interested not only in types of land cover, but also in the way the land was compositionally and spatially organized, so we can begin to assess the nature of human-environment interactions through the scale, pattern, and process paradigm that is emergent in ecology and other natural sciences.

Before proceeding, it is important to define certain terms. We are interested in the organization of land at the patch, class, and landscape levels. A patch is what makes up a landscape; it is defined as a discrete area of relatively homogeneous environmental conditions whose boundaries are distinguished by differences in environmental character from its surroundings. A class is a collection of patches of a similar cover type. A landscape is composed of a mosaic of patches and classes of similar and different types. Using available computer packages,[2] we created a set of pattern metrics measuring such variables as the fragmentation of the landscape and the size of the various classes. We are using these metrics to quantify the spatial and temporal changes in land-use/land-cover composition and pattern (or landscape structure) at the class and landscape levels as a signature of landscape form and function over time and space. For example, as part of our work on land use/land cover and migration, we hypothesized a correspondence between diversity of land use/land cover, as measured by pattern metrics such as number of patches within a fixed radius, and fragmentation of land ownership. In general, a larger number of patches would suggest a landscape where it is difficult to introduce new crops, especially if those crops require a large, contiguous parcel of land (typical of cassava growing in Nang Rong). Preliminary analysis suggests that land fragmentation encouraged out-migration of young adults during 1984-1994, as would be expected based on this line of argument (Rindfuss et al., 1996a).

The final component of our Nang Rong data is three sets of maps. The first set, and the one we have used most extensively, was obtained from the Thai Ministry of Defense. These thematic maps, produced in 1984 at a scale of 1:50,000, have been digitized to create base coverages of the study area, hydrography, transportation (including trails and foot paths), topographic contours and point elevations, village centers, market towns, district boundaries, and subdistrict health centers. In addition, a set of biophysical and geographical coverages, described below, has been derived through manipulation of the base coverages. A second set of maps, produced in 1992-1993, has also been digitized. These maps were produced for planning and administrative purposes.[3] A third set of maps (e.g., soils) has been obtained and also entered into the GIS.

Figure 6-2 provides an overview of the various remote sensing and population data sources used by the Nang Rong data laboratory. One inference that can clearly be drawn from this figure is that additional data sources can be readily incorporated. As time elapses, more data can be added to the right-hand side of the figure. Additional remotely sensed images can also be added for the time period already covered. Additional data layers can be added as well. For example, we are now in the process of obtaining aerial photographs for selected years dating back to the 1950s.

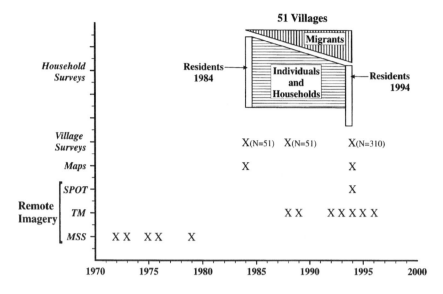

FIGURE 6-2   Nang Rong data: Sources and temporal coverage.

RESEARCH ISSUES

Several research issues have emerged as a result of our work in Nang Rong. These issues relate to the challenges of linking people and pixels, the availability of exogenous variables, and the scale of analysis.

Linking People and Pixels

A challenge in merging social science and remote sensing approaches is identifying congruent units of observation and developing appropriate linkages. Given a largely agricultural study area, we need to consider individuals, households, and villages on the social side; plots and territories on the spatial side; and ways of linking all of these units to each other and to pixels (as captured in the remotely sensed data). This challenge is illustrated in Figure 6-3. While conceptually straightforward, these links can be difficult to effect operationally. Certainly this is true for the Nang Rong study area, where people live in nucleated village settlements away from their fields and where households cultivate multiple noncontiguous agricultural plots.

Consider the comparison of settlement patterns in parts of Amazonia and Nang Rong, Thailand. In Amazonia, a common settlement pattern is for households to own and live on a single, large, coterminous plot of land. A stylized version of this pattern, characteristic of planned colonization and termed the "fishbone pattern" because of its appearance from space (Moran et al., 1994), is

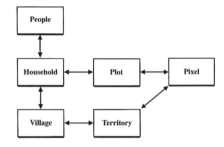

FIGURE 6-3   Units of observation and links among them.

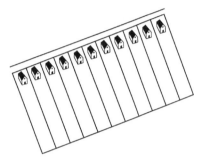

FIGURE 6-4   Fishbone pattern: Stylized relationship between place of residence and land ownership and use.

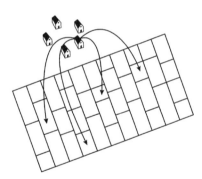

FIGURE 6-5   Place of residence and land ownership and use in Nang Rong.

shown as Figure 6-4. It is conceptually straightforward to link people, plots, and pixels given this pattern, although there are complications in practice, as indicated in the chapter by Moran and Brondizio in this volume. The settlement and farming pattern in Nang Rong departs from the stylized fishbone pattern in several important ways, as illustrated in Figure 6-5. Household residences are clustered in villages; the location of a household provides few clues about the location of the land farmed by that household. Further, households farm multiple plots (as many as 11), and these plots may be and often are scattered. If we had the spatial coordinates of all the plots farmed by all households in a village, as well as a clearly defined village territory, it would be possible to move among all the units of observation indicated in Figure 6-3. Since we do not have these spatial coordinates, we link at the village level instead.

The population surveys conducted in 1984, 1988, and 1994 at the household and village levels are represented in the GIS as discrete point locations mapped at the village centroid. Such a spatial representation is appropriate given the nuclear settlement pattern of Thai villages. Biophysical and geographical information derived through remote sensing techniques, however, represents space as a continuous surface summarized at the pixel level. Integration of social and environmental data therefore requires a transformation of space whereby a polygon representation is used to denote the pattern and variability of landscape conditions associated with discrete village locations. This transformation involves defining village boundaries, a complex issue in a region where political delineations change as village populations grow, and social, cultural, biophysical, and geographical parameters combine to influence the geometry and areal extent of village boundaries.

Our initial approach is fairly simple and involves generating radial buffers around the nuclear village centroids at distances of 2 and 3 km. This model is simple to implement and has particular relevance to the study area in that it (1) assumes travel distances from households to field plots appropriate to travel conditions during the 1984-1994 study period, which involved primarily walking; (2) provides for overlapping village boundaries associated with the density and location of villages; and (3) represents the nuclear village settlement concept, whereby residents disperse from village centers to engage in various land activities. Plate 6-2 (after page 150) shows an example of 3-km buffers for our 51 study villages overlaid on a 1993 TM land-cover classification. This figure makes it clear that villages may be competing with one another for available land, and we include measures of competition in our models (Entwisle et al., 1997b; Rindfuss et al., 1996b).

We are also beginning to explore other approaches for setting village boundaries, including Thiessen polygons, population-weighted Thiessen polygons, and the Triangulated Irregular Network (TIN). Such approaches produce nonoverlapping village boundaries and irregular boundary dimensions and orientations. Thiessen polygons maintain a one-to-one relationship between points and polygons. Any location within the defined polygon is closer, in Euclidean distance, to the associated point than to any of its neighboring points, and polygons vary in size inversely with point density. Plate 6-3 (after page 150) shows an example using a Thiessen polygon approach incorporating all villages in Nang Rong overlaid on a 1993 Landsat image.

In addition, doctoral research by Tom P. Evans noted earlier is examining the concept of distinct and fuzzy boundaries for representing territories of Thai villages. A fuzzy boundary can be created, for example, by land-ownership patterns, a product of the degree of land fragmentation around a village. Land fragmentation may develop as a consequence of intradistrict migration, land-inheritance practices, land-tenure patterns, and environmental variation. Fuzzy boundaries between villages suggest, in the context of Thai culture and society, the concept of functional boundaries that overlap in space and vary by social, cultural, biophysical, and geographical domains.

Availability of Exogenous Variables

We are interested in the interrelationships between population growth and change on the one hand and land-use and land-cover change on the other. While it is possible to examine associations, we also want to be able to examine causal relationships. The difficulty here is that the settlement of Nang Rong predates our data. Even at the earliest point in our time series, the land-cover pattern has been influenced by human behavior, and human behavior has been influenced by land-cover patterns. To draw causal inferences, one must be able to identify variables exogenous to the system of interrelationships. One such variable is elevation, which thus far in Nang Rong has not been influenced by human behavior or changes in land cover (Entwisle et al., 1997b).

The contour lines and point elevations on our 1:50,000 scale base maps have been scan-digitized and attributes attached, edge-matched, and processed to yield digital elevation models (DEMs) of the study area. Figure 6-6 shows the topographic inputs for the DEMs. The DEMs are being used to characterize the percentages of alluvial plain/lower terrace, middle/high terrace, and upland in the 2- and 3-km boundaries associated with each of the district villages.[4] The construction of DEMs is important for the characterization of topographic site and situation relationships associated with land-use/land-cover potentials (Entwisle et al., 1997b). For example, most recent agricultural production, occurring on low terraces and alluvial plains, has seen a substantial amount of the agricultural intensification related to paddy rice production.

In addition to obtaining elevation from the DEMs, we plan to work with several other derived measures to see if we can obtain exogenous measures of land suitability for various types of land cover and land use. One of the most common indices derived from DEMs, used to indicate potential wetness for a site, is the Topographic Convergence or Wetness Index (Beven and Kirby, 1979; Moore et al., 1991; Wolock and McCabe, 1995). This index uses calculations of slope and upslope contributing area (the area that drains through a given location) to generate a dimensionless index of potential wetness, which provides a simple but effective model used to characterize channels, gullies, and moisture sinks (Allen and Walsh, 1996; Townsend and Walsh, 1996). A digitized hydrography layer was used to support the calculation of the index. Topographic curvature was generated to integrate planform and profile curvature in order to portray localized topographic situations (Moore et al., 1991) within the soil moisture potential surface. The variable distinguishes slopes having a convex versus a

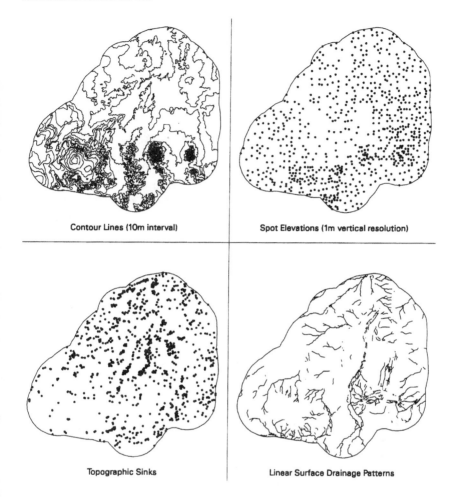

Contour Lines (10m interval)   Spot Elevations (1m vertical resolution)

Topographic Sinks   Linear Surface Drainage Patterns

FIGURE 6-6   Topographic data inputs for generation of digital elevation models, Nang Rong, Thailand.

concave slope. Profile curvature is the vertical or downslope concavity/convexity and portrays slope steepness. Slope curvature affects the acceleration and deceleration of gravity flows. Solar radiation affects the soil moisture, temperature regimes, and photosynthetic capacity of the landscape. The potential direct solar insolation at any location will be represented through grid locations throughout the study area, given DEM-derived elevation, slope angle, and slope aspect information. The procedures are modeled after Bonan (1989) and Montieth and Unsworth (1990). Shadowing by adjacent terrain is an important factor influencing potential insolation. Solar radiation potential is derived by integrating the effects of shadows, direct solar insolation, and insolation variability throughout the normal growing season within the study area. Values can be calculated for various time steps on an hourly, daily, and seasonal basis, and in future work we plan to explore the most appropriate time periodicity.

Scale of Analysis

The effects of spatial scale, pattern, and process are inherently involved in landscape studies. Processes shaping the landscape function only at certain ranges of spatial and temporal scales, and their effects vary with scale (Woodcock and Strahler, 1987; Nellis and Briggs, 1989). An important consideration in linking survey and image time series data is thus the scale of analysis. The spatial resolution of remote sensing systems is the pixel, whose dimensions vary with the sensor system and reconnaissance platform. As a consequence of the continuous nature of remotely sensed data organized within a raster data model, data aggregation approaches are available for altering the pixel dimension to a coarser grain size where appropriate. Scale-dependent analyses use such an approach to define the range of spatial scales of remotely sensed data and continuous GIS coverage that may be highly autocorrelated in an effort to interrelate scale, pattern, and process.[5] Integration of data collected from different satellite systems (e.g., SPOT MX, Landsat TM, the National Oceanic and Atmospheric Administration's [NOAA] Advanced Very High Resolution Radiometer [AVHRR]) offers an additional approach to spatial resolution through information scaling. While population data can be transformed from discrete point locations to a continuous surface through spatial interpolation and spread approaches, natural population units (e.g., household, village, district) may be more suitable. The summarizing of satellite data at the village level is subject to difficulties related to the setting of village boundaries, whereas the use of agricultural plots as the unit of measure is subject to the spatial resolution limitations of the satellite systems and the constraints of the population survey that make it difficult to link households to geographic lot locations.

The temporal dimension of remotely sensed data is achieved by acquiring images for single time slices that are combined into a time series. Population-based life histories and retrospective analyses based on population surveys offer a distinctly different approach for representing the temporal dimension as seen through the creation of an image time series.

While questions of information scaling within social and biophysical systems need to be addressed more effectively, the scaling of social and biophysical relationships across temporal and spatial domains and within similar and different environments offers unique research challenges. In our research, we have defined a scale continuum for the population, environment, and geographical domains that includes, respectively, (1) household, village, and village clusters; (2) field plot, terrain strata, village, and watershed; and (3) pixel, aggregated pixel, village, terrain strata, and watershed. The temporal range extends from the 1970s to the present, with particular emphasis on 1984, 1988, and 1994. The temporal scale of analysis is the season, year, and decade. Research is ongoing at various scales along the continuum, and other research is planned. A challenge is to link population-environment data at various scales, define important variables along the scale continuum, scale the variables from one resolution to another, and assess the important processes and relationships across domains and across spatial and temporal scales.

CONCLUSIONS

In conducting the research described above, our research team has benefited from expertise across a range of substantive and methodological specialties. In our case, long-standing research programs in social and biophysical topics have been integrated to address population-environment questions through a synergistic relationship among sociology, demography, geography, and spatial techniques. While the exchange of language and concepts among disciplines is still occurring, team members assumed leadership roles in various substantive and methodological areas early in our research relationship to move quickly up the learning curve. Literature exchanges, processing discussions, conceptual dialogue, location at the same university, standing in one's scientific community, similarity of research styles and modes of work, and early success with funding agencies contributed to our collaborative efforts. Other advantages of our research collaboration included team proficiencies in a broad-based set of research methods, as well as an appropriate computer infrastructure to support remote sensing image processing, GIS, spatial analysis, GPS technology, and statistical modeling. The availability of population survey data collected for the study area in 1984, 1988, and 1994 was another essential ingredient of both our start-up efforts in linking spatial and social data and our sustained program of inquiry. The survey data and its intensive preprocessing at IPSR, Mahidol University, and the CPC provided the temporal and longitudinal depth for our ongoing analyses. Finally, especially in the early years of the project, we benefited from the willingness of funding agencies and program administrators to take a risk on an unproven research plan. Consider these comments by a reviewer of one of our earliest proposals:

> This research proposal is complex and ambitious....It is highly innovative and interdisciplinary. This presents pitfalls as well as exciting opportunities....It is difficult to assess the potential value of the research; it is an experiment, and we won't know until it's been tried....My reaction is to forgive the pitfalls and to embrace the experiment as plausible and worthwhile (but risky). I suspect that serious efforts to incorporate environmental factors into social science will require spatial and interdisciplinary research.

Fortunately, the funding agency followed the recommendation of this reviewer and provided crucial start-up money. This sort of vision on the part of funding agencies and program administrators is still needed.

Remotely sensed data provide a valuable landscape perspective that is substantially enhanced through the availability of corresponding social data. Satellite time series offer the best opportunity for landscape analyses over time and space. The use of remotely sensed data for international sites, however, is problematic in that archived data are often limited, multitemporal data for examining interannual or interseasonal variations are difficult to assemble and acquire, and only a few commonly available satellite systems offer alternate views of the landscape. Finally, while remote sensing systems and data accessibility could be enhanced, population surveys could also benefit from the routine geocoding of landscape features (e.g., households and village centroids) as part of the survey, the formulation of survey questions that directly examine spatial pattern and geographic location (e.g., spatial networks and field plot positions), and the examination of population-environment issues across a range of spatial and temporal scales.

NOTES

1 The planning for the 1994 survey began in 1990, well before Aphichat Chamratrithirong, Barbara Entwisle and Ronald R. Rindfuss were aware of the potential for linking the social data to satellite data.

2 Turner's (1990) SPAN (Spatial Analysis) algorithm was one of the first widely distributed landscape pattern programs. Users of the Geographic Resource Analysis System (GRASS) use a module developed by Baker and Cai (1992) for examining multiscale landscape patterns in the "r.le" supplemental programs. DeCola and Montagne (1993) have published results from the PYRAMID system, a package designed to analyze multiscale spatial structure in raster data sets, and McGarigal and Marks (1993) developed FRAGSTATS, a widely used package of composition and pattern metrics. Other recently developed packages for assessing composition and pattern include APACK, a free-standing package for spatial analysis of raster data; DISPLAY, which calculates landscape structure through a set of metrics that can be processed in batch mode and implemented sequentially; RULE, which assesses landscape organization according to defined neighborhood dimensions; and HISA (Habitat Island Spatial Analysis), used to calculate a host of landscape metrics from digital land-cover maps.

3 The 1984 maps showed nucleated settlement patterns as "villages," irrespective of the number of households in the village. The 1992-1993 planning maps showed administrative village locations, based on an administrative system that aims for villages of approximately 100 households. Needless to say, with two different concepts behind the mapping procedures, the two maps do not agree, even taking into account the time differences between them. At the time of this writing, field work is in progress that will result in our being able to move back and forth between these two conceptually distinct types of villages.

4 The TOPOGRID module in ARC/INFO was used to interpolate the DEM from the scanned contour data. TOPOGRID generates a DEM from contour lines, point elevation data, and hydrography information using an iterative finite difference interpolation technique developed by Hutchinson (1989). The algorithm interpolates elevation values iteratively using a thin plate spline, thereby developing an optimal flow model that maintains the integrity of the input data while simultaneously ensuring surface continuity (Hutchinson, 1993). The algorithm is optimized by using digital hydrography data to constrain TOPOGRID to identifying the appropriate low elevations along drainages.

5 Semivariograms and fractals will be used to assess the nature and degree of the autocorrelation among landscape features at different spatial scales. Walsh has explored the scale dependence among elevation, slope angle, and slope aspect, derived from GIS-based DEMs, and plant biomass, defined through satellite digital data in an alpine environment in Montana (Bian and Walsh, 1993). Bian and Walsh addressed (1) the effective range of spatial scales within which plant biomass and terrain variables were spatially dependent; (2) optimum spatial scales for representing terrain and plant relationships; and (3) the degree of spatial dependence of these relationships. Walsh et al. (1997) report that the spatial organization of the Normalized Difference Vegetation Index (NDVI) at finer scales is an important indicator of the NDVI at coarser scales.

REFERENCES

Allen, T.R., and S.J. Walsh
  1996 Spatial and compositional pattern of alpine treeline. Glacier National Park, Montana. *Photogrammetric Engineering and Remote Sensing* (in press).
Baker, W.L., and Y. Cai
  1992 The r.le programs for multiscale analysis of landscape structure using the GRASS geographical information system. *Landscape Ecology* 7(4):291-302.
Beven, K.J., and M.J. Kirby
  1979 A physically-based variable contributing area model of basin hydrology. *Hydrological Sciences Bulletin* 24(1):43-69.
Bian, L., and S.J. Walsh
  1993 Scale dependence of vegetation and topography in a mountainous environment of Montana. *Professional Geographer* 45(1):1-11.
Bilsborrow, R.E., T.M. McDevitt, S.A. Kossoudji, and R. Fuller
  1987 The impact of origin community characteristics on rural-urban out-migration in a developing country. *Demography* 24:191-210.
Bilsborrow, R.E., A.S. Oberai, and G. Standing
  1984 *Migration Surveys in Low Income Countries: Guidelines for Survey and Questionnaire Design.* London, England: Croom Helm.
Bonan, G.B.
  1989 A computer model of the solar radiation, soil moisture, and soil thermal regimes in boreal forests. *Ecological Modelling* 45:275-306.
Caplow, T., H.M. Bahr, et al.
  1982 *Middletown Families: Fifty Years of Change and Continuity.* Minneapolis: University of Minnesota Press.
Clausen, J.A.
  1972 The life course of individuals. Pp. 457-514 in *Aging and Society. Vol. 3: A Sociology of Age Stratification,* M.W. Riley, M. Johnson, and A. Foner, eds. New York: Sage.
Curran, S.R.
  1994 Household resources and opportunities: The distribution of education and migration in rural Thailand. Ph.D. dissertation, Department of Sociology, University of North Carolina.
Davis, K.
  1963 The theory of change and response in modern demographic history. *Population Index* 29:345-366.
DeCola, L., and N. Montagne
  1993 The PYRAMID system for multiscale raster analysis. *Computers and Geosciences* 19(10):1393-1404.

Donner, W.
   1978   *The Five Faces of Thailand: An Economic Geography.* London, England: Hurst Publishing.
Elder, G.H., Jr.
   1985   Perspectives on the life course. Pp. 23-49 in *Life Course Dynamics: Trajectories and Transitions, 1968-1980,* G.H. Elder, Jr., ed. Ithaca, N.Y.: Cornell University Press.
Elder, G.H., Jr.
   1991   Making the best of life: Perspectives on lives, times, and aging. *Generations* 15:12-17.
Entwisle, B., R.R. Rindfuss, S.J. Walsh, T.P. Evans, and S.R. Curran
   1997a Geographic information systems, spatial network analysis, and contraceptive choice. *Demography* 34:171-187.
Entwisle, B., S.J. Walsh, and R.R. Rindfuss
   1997b Population Growth and the Extensification of Agriculture in Nang Rong, Thailand. Paper presented at the annual meetings of the Population Association of America, Washington, D.C.
Entwisle, B., R.R. Rindfuss, D.K. Guilkey, A. Chamratrithirong, S.R. Curran, and Y. Sawangdee
   1996   Community and contraceptive choice in rural Thailand: A case study of Nang Rong. *Demography* 33(1):1-11.
Entwisle, B., and W.M. Mason
   1985   The multilevel effects of socioeconomic development and family planning programs on children ever born. *American Journal of Sociology* 91:616-649.
Findley, S.
   1987   *Rural Development and Migration: A Study of Family Choices in the Philippines.* Boulder, Colo.: Westview Press.
Freeman, L.C.
   1979   Centrality in social networks: I. Conceptual clarification. *Social Networks* 1:215-239.
Fukui, H.
   1993   *Food and Population in a Northeast Thai Village.* Monographs of the Center for Southeast Asian Studies, Kyoto University. English-Language Series, No. 19. Honolulu: University of Hawaii Press.
Hutchinson, M.F.
   1989   A new procedure for griding elevation and stream line data with automatic removal of spurious pits. *Journal of Hydrology* 106:211-232.
   1993   Development of a continent-wide DEM with applications to terrain and climate analysis. Pp. 393-399 in *Environmental Modeling with GIS,* M.F. Goodchild, B.O. Parks and L.T. Steyaert, eds. New York: Oxford University Press.
Institute for Population and Social Research (IPSR)
   1984   *A Demographic, Socioeconomic, and Health Profile of a Rural Community in Nang Rong.* Bangkok, Thailand: Institute for Population and Social Research, Mahidol University.
Limpinuntana, V., A. Patanothai, T. Charaoenwatana, G. Conway, M. Seetisarn, and K. Rerkasem
   1982   *An Agroecosystem Analysis of Northeastern Thailand.* Khon Khaen, Thailand: Faculty of Agriculture, Khon Khaen University.
Massey, D.S., and F. Garcia Espana
   1987   The social process of international migration. *Science* 237:733-738.
McGarigal, K., and B.J. Marks
   1993   FRAGSTATS: Spatial Pattern Analysis for Quantifying Landscape. Version 1.0. Forest Science Department, Oregon State University.
Montieth, J.L., and M.H. Unsworth
   1990   *Principles of Environmental Physics.* London, England: Edward Arnold.
Moore, I.D., R.B. Grayson, A.R. Landson
   1991   Digital terrain modelling: A review of hydrological, geomorphological, and biological applications. *Hydrological Processes* 5:3-30.

Moran, E. F., E. Brondizio, and P. Mausel
1994   Integrating Amazonian vegetation, land-use, and satellite data. *BioScience* 44:329-338.
Nellis, M.D., and J.M. Briggs
1989   The effect of spatial scale on Konza landscape classification using textural analysis. *Landscape Ecology* 2(2):93-100.
Panayotou, T., and S. Sungsuwan
1989   An Econometric Study of the Causes of Tropical Deforestation: The Case of Northeast Thailand. Harvard Institute for International Development Discussion Paper No. 284, Harvard University.
Parnwell, M.J.G.
1988   Rural poverty, development and the environment: The case of North-East Thailand. *Journal of Biogeography* 15:199-208.
Pendleton, R.L.
1963   *Thailand: Aspects of Landscape and Life.* New York: Dwell, Sloan, and Pearce.
Phantumvanit, D., and K.S. Sathirathai
1988   Thailand: Degradation and development in a resource rich land. *Environment* 30:11-15, 30-32.
Phillips, J. F., R. Simmons, M.A. Koenig, and J. Chakraborty
1988   Determinants of reproductive change in a traditional society: Evidence from Matlab, Bangladesh. *Studies in Family Planning* 19:313-334.
Rindfuss, R.R., S.J. Walsh, and B. Entwisle
1996a   Land Use, Competition, and Migration. Paper presented at the Population Association of America Meeting, New Orleans, La.
Rindfuss, R.R., D.K. Guilkey, B. Entwisle, A. Chamratrithirong, and Y. Sawangdee
1996b   The family building life course and contraceptive use: Nang Rong, Thailand. *Population Research and Policy Review* 15:341-368.
Rogers, E.M.
1983   *Diffusion of Innovation.* New York: Free Press.
Sawangdee, Yothin
1995   Multi-Level Analysis of Factors Influencing An Adult Temporary Migrant's Destination: A Case Study From Nang Rong, Thailand. Unpublished M.A. thesis, University of North Carolina.
Siamwalla, A., S. Setboonsarng, and D. Patamasiriwat
1990   Agriculture. Pp. 1-117 in *The Thai Economy in Transition*, P.G. War, ed. New York: Cambridge University Press.
Tambunlertchai, S.
1990   *A Profile of Provincial Industries.* Bangkok: Thailand Development Research Institute Foundation.
Townsend, P.A., and S.J. Walsh
1996   Spatial variability of a wetness model to soil parameter estimation approaches. *Earth Surface Processes and Landforms* 21(4):307-326.
Walsh, S.J., T.P. Evans, W.F. Welsh, R. R. Rindfuss, and B. Entwisle
in   Population and environmental characteristics associated with village boundaries and
press   landuse/landcover patterns in Nang Rong district, Thailand. *Proceedings of Pecora 13, Symposium on Human Interactions with the Environment: Perspectives from Space.* Bethesda, Md.: American Society for Phtogrammmetry and Remote Sensing.
Walsh, S.J., A. Moody, T.R. Allen, and D.G. Brown
1997   Scale dependence of NDVI and its relationship to mountainous terrain. Pp. 27-55 in *Scaling in Remote Sensing and GIS* D.A. Quattrochi and M.F. Goodchild, eds., Boca Raton, Fla: CRC/Lewis Publishers.

Wolock, D.M., and G.J. McCabe, Jr.
 1995   Comparison of single and multiple flow direction algorithms for computing topographic parameters in TOPMODEL. *Water Resources Research* 31:1315-1324.
Woodcock, C.E., and A.H. Strahler
 1987   The factor of scale in remote sensing. *Remote Sensing of Environment* 21:311-332.

7

Validating Prehistoric and Current Social Phenomena upon the Landscape of the Peten, Guatemala

Thomas L. Sever

Remotely sensed data from airborne and satellite sensors have been used to identify archeological features, such as roadways, temples, cisterns, and agricultural areas. This information is critical in research on the Maya to help answer questions regarding subsistence, settlement patterns, population densities, societal structure, communication, and transportation. These issues in turn may well relate to perhaps the most intriguing question of all: the reason for the Mayan collapse.

From the perspective of space we can also view the effects of human activity upon the landscape. Satellite images of the Peten region of Guatemala allow us to see where, when, and how rapidly that landscape is changing. Satellite data can be used to provide quantifiable evidence for depletion rates and trends of deforestation, identify potential points of conflict, and create predictive models for the future. The data can also be combined with ground-truth information to help us better understand the social issues involved and develop an effective strategy for balancing the challenges of population increase, sustainability, and conservation.

Because of unanticipated preliminary results, this project evolved from a small-scale study to a regional analysis. Originally, the project was to conduct an archeological and environmental analysis of a 1 × 1 degree study area along the Usumacinta river valley between Guatemala and Mexico. This work was to be performed in response to potential hydroelectrical projects and large-scale oil explorations that threatened the area. When the first Landsat Thematic Mapper (TM) imagery of the area was processed, it revealed the contrast between the tilled landscape of Mexico and the standing rainforest of Guatemala (Plate 7-1,

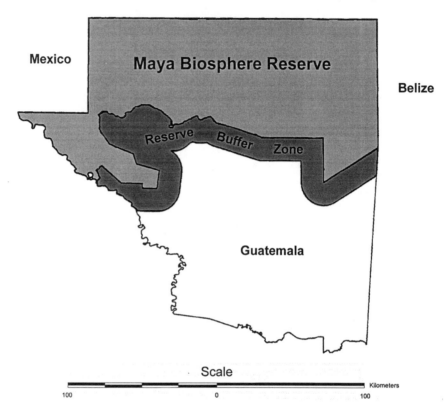

FIGURE 7-1   Location of the Maya Biosphere Reserve and buffer zone in the Peten district of northern Guatemala.

after page 150). This image received national and international attention with its appearance in the October 1989 issue of *National Geographic* magazine. It was also taken before the Guatemala Congress by President Vinicio Cerezo, which resulted in the establishment and protection of the Maya Biosphere Reserve (MBR) in 1990. This image also led to an agreement between the National Aeronautics and Space Administration (NASA) and the Central American Commission on Environment and Development (CCAD) to conduct joint research as part of the Mission to Planet Earth Program. As a result of these developments, the original study area was expanded to include both the Peten region of Northern Guatemala and the MBR, the protected area within the Peten's boundaries (see Figure 7-1).

Initially, the primary focus of this investigation was on the use of remote sensing technology for the identification of unrecorded archeological features. However, the utility of the remote sensing data for monitoring of current deforestation activities was also apparent to conservationists, scientists, managers, and politicians. Consequently, two avenues of research—archeological inventory and deforestation—have proceeded in parallel throughout the course of the project. These two areas of research are discussed separately here, although both are related to the social sciences.

Using airborne and satellite remote sensing data, the project has identified prehistoric archeological features, such as roadways, temples, and agricultural areas, as well as documented a time-lapse sequence of deforestation from 1986 to the present (Sader et al., 1996). The clearing of tropical forests is one of the primary global environmental issues. Deforestation leads to three unintended impacts: loss of cultural diversity, loss of biodiversity, and loss of carbon storage capacity. Satellite imagery has proven to be one of the best techniques for monitoring forest clearing, shifting cultivation, and land-use conversion patterns (Sader, 1995; Sader et al., 1994). While remotely sensed data provide quantifiable information about forest and land-cover changes, they do not explain the anthropogenic causes for those changes. Hence ground-truth information is required both to validate the satellite imagery and to supplement it with information from local harvesters, farmers, and ranchers regarding the criteria they employ for land-use conversion.

Our research team consists of James Nations, an anthropologist and ecologist at Conservation International; Santiago Billy, a Guatemalan national and conservationist for Conservation International; Frank Miller, a forester at Mississippi State University; Daniel Lee, a geographic information system (GIS) expert at GeoTek; and the author, a remote sensing specialist and archeologist at NASA. This interdisciplinary team has been conducting field research in the Peten for 10 years to verify the results of our computer analysis. The primary focus of our research is environmental inventory and the detection of archeological features within the Peten. During the course of our work we have collaborated with many other researchers, particularly Steven Sader, University of Maine, in deforestation studies and Patrick Culbert, University of Arizona, in archeological research.

BACKGROUND

The Peten, in northern Guatemala, covers 36,000 km^2—a third of Guatemala's land mass. The wetlands and intact tropical forest of the Peten represent one of the world's richest areas of biological diversity. The ecosystem contains over 800 species of trees; 500 species of birds; and large populations of mammals, including monkeys, jaguars, and tapirs. The area also contains some of the most prehistorically significant Mayan archeological sites in Latin America. Since deforestation activities also include the destruction of archeological sites, conservation of these ruins is synonymous with forest protection. Only a few indigenous Mayan descendants still live in the Peten, although the population of inhabitants is increasing rapidly as a result of migration and settlement. Until 1970, nearly 90 percent of the Peten remained forested. Today over half of the forest has been cut, and if deforestation continues at its present rate, only 2 percent of the Peten's forest will remain by the year 2010 (Canteo, 1996).

The MBR lies within the Peten and covers 1.6 million hectares (ha), nearly 40 percent of the Peten and 14 percent of Guatemala. It is the largest protected area in of the Maya Tropical Forest, which stretches from Eastern Mexico to Belize, and as such is critical to regional conservation efforts. Deforestation rates in the area are accelerating as a result of increasing human migration. Over the past 20 years, the human population in the Peten has increased from 20,000 to more than 300,000 (Stuart, 1991). Today, the Peten has a human population density of 26 inhabitants per square mile (J. Nations, personal communication); this figure contrasts significantly with the ancient Mayan population densities of 2,600 persons per square mile in the center and 500-1,300 per square mile in the more rural areas (Rice, 1991).

New roads and pipelines being constructed in the Peten by logging and petroleum companies serve as conduits for human migration (Plate 7-2, after page 150). Peasant farmers follow these corridors to clear the forest and establish maize-based agricultural plots (milpas). Slash-and-burn cultivation remains the predominant agricultural system in the Peten today. Soil fertility declines after 2 or 3 years of cultivation, and the milpas are abandoned (Lundell, 1937) as the farmers move to new areas. In addition, a major expansion of the Q'eqchi' Maya population has occurred in the southern Peten. The Q'eqchi', numbering over 400,000, show remarkable abilities in adapting to new environments and learning new technologies (Cahuec and Richards, 1994). Cattle pastures are becoming increasingly present in the Peten, although they are not yet dominant as in other regions of Central America.

The ancient Maya far outgrew the carrying capacity of slash-and-burn agriculture, which is only 200-250 persons per square mile in this region (Rice, 1991). The environmental effects of the current deforestation and settlement in the area are not known at this time. What is known is that monocultivation and cattle ranching are replacing the traditional adaptive system of the past. The forests of the Peten were nearly destroyed 1000 years ago by the ancient Maya, who, after centuries of successful adaptation, had finally overused their resources. Current inhabitants are threatening to do the same thing today in a shorter time period with a lesser population.

ARCHEOLOGICAL RESEARCH

The Classic Mayan period of the southern lowlands lasted from 250 to 900 A.D., when it collapsed in unparalleled fashion. That some disaster had befallen the Classic Maya became clear in the early days of Mayan studies when it was learned that the dating and inscription of monuments had ceased during the ninth century. The seventh and eighth centuries were a golden age for the Maya. A population of millions achieved new peaks in construction. Stone monuments proclaimed the glory of the rulers in decorative masterpieces. Between 830 and 930 A.D., however, populations of both elites and commoners in the Mayan lowlands declined by two-thirds. By the Postclassic Period (after 930 A.D.), only a few scattered houses remained (Culbert, 1993).

The Mayan collapse may well be the greatest demographic disaster in human history. It occurred in the region of the Mayan lowlands. There was no exodus outward because the peripheral populations remained stable. What caused this disaster is not known. Potential explanations include internal conflict and warfare, disease, climatic change, overpopulation, outside invasion, peasant rebellion, soil nutrient depletion, and failure of the agricultural system (Culbert, 1974, 1993; Hammond, 1990). Rather than having a single explanation, however, the cataclysm may have had multiple causes. What is known is that most of the trees in the region had been cut down by the ninth century. Although there is recent evidence for a drought beginning around 800 A.D. (Sabloff, 1995; Curtis et al., 1996), the question emerges of why the Maya had survived previous droughts, not to mention other setbacks. Understanding how the Maya managed the landscape so successfully for centuries and what the eventual effect was, if any, of their deforestation activities may provide an important lesson for the future of today's inhabitants.

TM imagery reveals Mayan causeways that often cannot be seen from ground level. As part of the theoretical framework of landscape archeology, roadways reflect the interplay among technology, environment, social structure, and the values of a culture (Trombold, 1991). In this case, the causeways are the tangible evidence of the Maya's structural organization across geographic space. The Mayan causeways are formal routes. Formal routes reflect elements of planning and purposeful construction, including labor in construction, engineering, maintenance, and an organizational structure that oversees the system. They tend to be straight in nature and are engineered to overcome natural obstacles in order to improve transportation and communication. They contrast with informal routes, such as paths, which show little if any evidence of construction, are irregular in design, and tend to avoid natural obstacles. From a remote sensing perspective, the Mayan roadways tell us how the cities were connected and to some degree suggest the level of sophistication of the city responsible for their construction and maintenance.

Several analytical techniques have been employed to identify Mayan causeways in the TM imagery. Laplacian and Gaussian filtering techniques were used to isolate linear and curvilinear features in the imagery. These techniques have been employed successfully to detect Anasazi roads in New Mexico (Sever, 1990) and prehistoric footpaths in Costa Rica (Sheets and Sever, 1988). In addition, a number of different ratios and other transformations, such as the Normalized Difference Vegetation Index (NDVI), principal component analysis, and a modified version of the Kauth-Thomas (KT) transformation (MKT) were used to isolate the causeway features. MKT was derived as follows:

- The KT tasseled cap transformation was performed using standard coefficients.
- Bands 1 and 2 of the KT transformation were ratioed using Band 1 - (Band 2/Band 1 + Band 2). In this format, the central processing unit (CPU) order of operations is to perform the division first and then subtract, yielding a different value from that of the NDVI calculation. These values are assigned to the red video color gun.
- Bands 4 and 5 were ratioed as 5/4 and assigned to the green video gun.
- Bands 5 and 7 were ratioed as 5/7 and assigned to the blue video gun.

The combination of these techniques made it possible to identify various roadway segments. For instance, as the roads cross the seasonally flooded swamps (bajos), they support a more lush vegetation than the surrounding plant community. Since the TM data are collected near the end of the dry season (April-May), the dryness of the roads in the elevated areas can be detected, as can the wetness in some of the lower areas. The most distinguishing characteristic, however, is the straightness or linearity of these features. The karst topography of the landscape also reveals linear geologic faults, fractures, and drainages that can be confused with causeways. Ground-truth reconnaissance can often resolve this issue visually. In some cases, a ground survey is not sufficient, and excavation must be used to distinguish the features conclusively. It should be mentioned that some of the techniques used to identify roadways also reveal temple structures and man-made water storage areas (see Figure 7-2). The 30-m resolution of the satellite imagery precludes detection of canals, but airborne 5-m Thermal Infrared Multispectral Scanner (TIMS) and Calibrated Airborne Multispectral Scanner (CAMS) data, as well as simultaneously acquired color infrared (CIR) photographs, have been used to identify potential canals and field systems. Although these features have not been verified to date, their existence is supported by the discovery of canal features in nearby study areas. The problem is that the 5-m data were not acquired in the specific bajo areas where the canals were discovered through excavation. As discussed below, the detection of these canals will provide insight into the use of the bajos by the Maya. The CIR photography was also used to confirm Mayan causeways (Figure 7-3).

One of the leading debates in Mayan archeology concerns whether the lowland wetlands (bajos) of the Peten were or could be used for agriculture (Pope and Dahlin, 1989, 1993; Adams et al., 1990; Turner, 1993). The bajos represent 40 percent of the land surface of the Peten, and it seems reasonable to assume that these areas were farmed by the ancient Maya. As noted earlier, slash-and-burn technology, still used by the inhabitants today, can support only 200-250 persons per square mile in the region. This level had been reached by the Maya by 300 A.D., far before their population peak in the Late Classic period (600-830 A.D.), when population densities ranged from 500 to 2,600 persons per square mile (Rice, 1991).

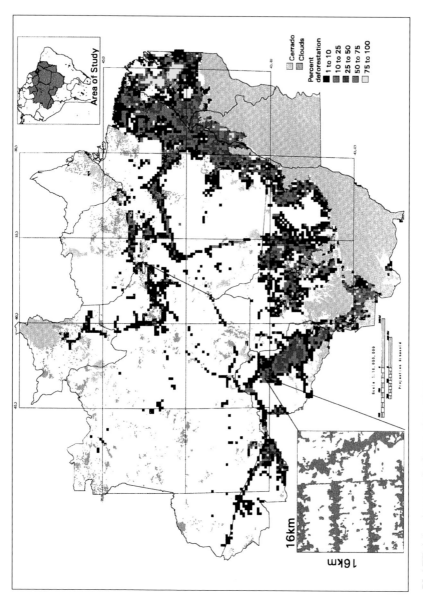

PLATE 4-1 Amazon Deforestation, 1986.

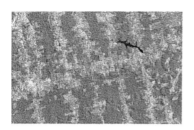

1 Altamira, Pará, Xingu Basin

- Landsat TM (226/62), dates 1985, 87,88,91,92

- Vegetation inventory: 18 Secondary Succession; 2 Upland forest (lianna and dense)

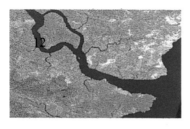

2 P. de Pedras, Pará, Marajó I.

- Landsat TM (224/61), dates 1984, 85, 87, 88, 91

- Vegetation inventory: 8 Secondary Succession; 2 Upland forests; 2 Fl. forest; 10 Açai agroforestry; 4 Savannas

3 Igarapé-Açú, Pará, Bragantina

- Landsat TM (223/61), dates 1984, 91, 95

- Vegetation inventory: 16 Secondary Succession; 2 Igapo forest; 1 Upland forest

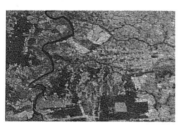

4 Tomé-Açú, Pará

- Landsat TM (223/61), dates 1984, 91, 95

- Vegetation inventory: 13 Secondary Succession; 1 Upland forest; 5 Agroforestry

5 Yapú, Vaupes, Colombia

- Landsat TM (004/59; 004/60), dates 1984, 93

- Vegetation inventory; 5 Secondary Succession; 1 Upland forest; 1 Sabanna alta; 1 Sabanna baixa (campinarana)

PLATE 5-1 Secondary succession in Amazonia: Thematic Mapper (TM) images of five research sites.

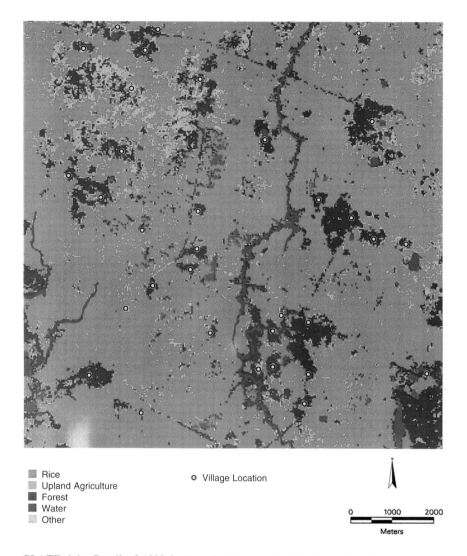

Rice
Upland Agriculture
Forest
Water
Other

o Village Location

0   1000   2000
Meters

PLATE 6-1   Detail of 1993 land-use/land-cover classification with village locations, Nang Rong, Thailand. SOURCE: Landsat TM Imagery.

Rice
Upland Agriculture
Forest
Water
Other

○ Survey Village Location
N 3-Km Radius Buffer

0   5   10
Kilometers

PLATE 6-2   1993 land-use/land-cover with survey villages and 3 km buffers, Nang Rong, Thailand. SOURCE: Landsat TM Classification.

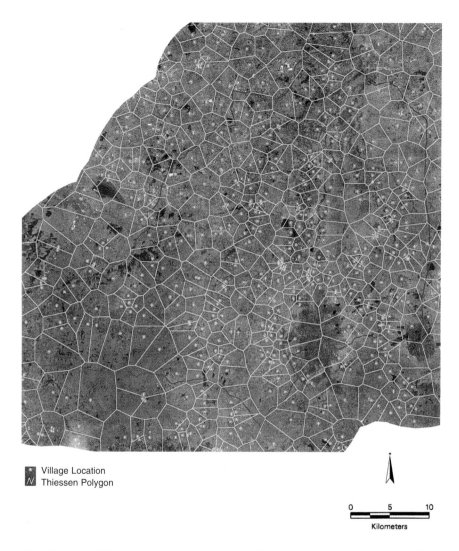

Village Location
Thiessen Polygon

0   5   10
Kilometers

PLATE 6-3   1993 raw Landsat TM (4,3,2) and Thiessen polygons, Nang Rong, Thailand.

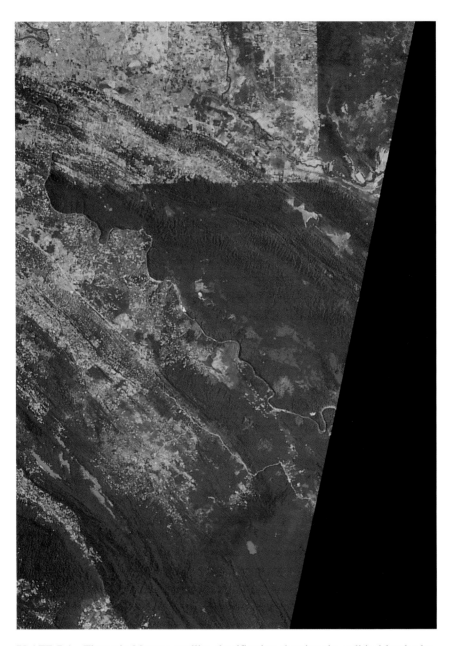

PLATE 7-1   Thematic Mapper satellite classification showing the political border between Mexico and Guatemala. This image reveals the impact of high rural population on
the rain forest. The dark green area represents Guatemala's sparsely populated Peten
district as it stands in contrast to the stripped and tilled landscape of Mexico.

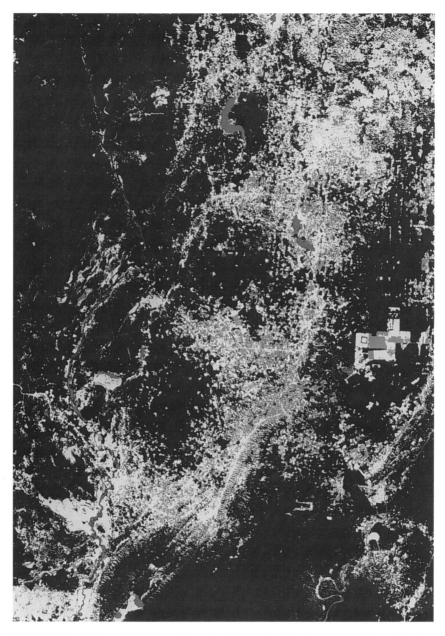

PLATE 7-2 Color-coded change-detection image showing the pattern of accelerating deforestation between 1986 and 1995. In 1986, only a single road extended through this part of the western Peten to the town of Naranjo, near the Mexico-Guatemala border, part of which can be seen in the lower left. In this image, blue represents areas cut down during 1986-1990, magenta during 1990-1993, and yellow during 1993-1995.

PLATE 7-3 Thematic Mapper satellite classification separating the bajos (seen as magenta) from the remainder of the rain forest (seen as green). Our current research is attempting to isolate various types of bajos, which will assist in answering questions about ancient Mayan farming, as well as help identify optimum areas for farming today.

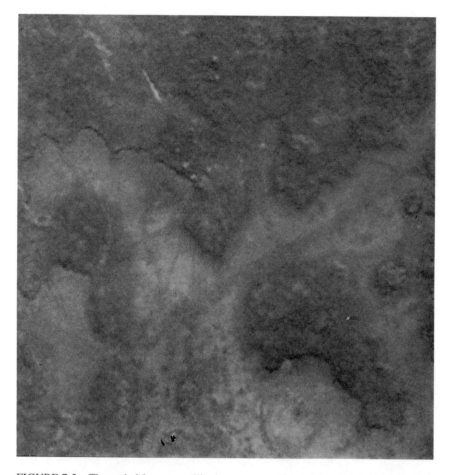

FIGURE 7-2   Thematic Mapper satellite image created by ratioing Band 4 with Band 3. The light spots represent ancient Mayan temple structures beneath the forest canopy at the archeological site of Mirador, while the dark lines represent Mayan causeways and natural geological features. Ground reconnaissance and sometimes excavation are required to separate the cultural from the natural features.

Currently, our research team has joined forces with Patrick Culbert of the University of Arizona to address this issue. Both supervised and unsupervised classification techniques (Kelley, 1983) have been used to classify satellite images and identify the bajo areas (Plate 7-3, on facing page). Three different types of bajos have been identified, in addition to seven additional land-cover classes, such as high forest, transition forest, and water. Local informants in the Peten, however, have identified to Culbert seven different types of bajos, each with

FIGURE 7-3 Photograph based on color infrared (CIR) image of the bajo to the immediate west of Mirador. The linear feature is a Mayan causeway. The causeways are artifacts from the past and assist in regional analysis since they provide insight into how Mayan cities were connected politically and economically.

distinctive characteristics. Of these seven types, two or three remain sufficiently moist in the dry season to sustain agriculture. This information is pertinent today, and may provide insight into early Mayan farming practices. Currently, we are applying supervised classification techniques using information from two transects as the training samples. The transect data come from field data collected by Culbert and his research team during their May 1995 field season. They conducted two transects in the Bajo la Justa, Peten—one a trail of approximately 7 km, the other along a road of approximately 17 km. They were assisted in the vegetation survey by a native (Don Felipe Lanza) employed by the Forestry Division of the Tikal National Park.

During this field reconnaissance they learned that there are two primary classifications of bajo associations: palm and scrub. Within each of these two major types are "sections characterized by a predominance of particular species which provide subtypes such as 'escobal', 'corozal', and 'botonal' within the palm bajo, and 'jimbal', 'tintal', 'navajuelal' and 'huechal' within the scrub bajo" (Culbert et al., 1995:3). Additionally, there are three distinctions that relate to ground-surface characteristics and inundation: "bajo plano," which has a flat surface and no noticeable slope; "bajo borbolal," which has an undulating surface; and "bajo inundable," which is at an elevation where seasonal flooding occurs and seems to be characterized by scrub vegetation. Lanza also informed Culbert and his team that palm bajo is excellent for milpa and is one of the environments of choice among native Peteneros. Along both transects the team recorded vegetation according to Lanza's classification and gathered Global Positioning System (GPS) readings.

This ground-truth information has been overlaid on the TM satellite data, and various classifications have been developed. In addition, we will acquire standard-beam (30-m) Radarsat C-band Synthetic Aperture Radar (SAR) data, which will then be merged with the TM data. This merging of TM/SAR data should provide the best opportunity for distinguishing class types. The various classification schemes will be tested during the 1998 field season to determine which technique is the most accurate.

In the process of searching for the best satellite band combinations, ratios, and other transformations for discriminating sacbes and bajos, it was noted that these techniques also discriminated elevated "islands" of upland within the bajos. The best technique for discriminating such islands employed the MKT using TM bands 3, 4, and 5. Culbert's 1995 field season led to key findings with regard to these islands:

> A very important fact about the location of Maya sites in relation to bajos has become clear...our work suggests that there was occupation on almost every area of slightly higher land where patches of high forests occur as "islands" within bajo. We call these areas of occupation "bajo communities" and they will be one of the foci of our future project in the Bajo la Justa (Culbert et al., 1995:4).

The discovery of these occupied islands within the bajo system raises many research questions related to chronology, settlement patterns, and artifactual assemblages. Do these islands represent a long-term occupation, or are they the result of a sudden, brief occupation? Are the bajo communities similar to or different from other communities, and do they operate under a similar or different organizational structure? The answers to these questions are critical in understanding the full adaptation of the Maya to their rainforest environment, and mapping these islands is the first step toward finding these answers.

Recently, our research team participated in the identification of an unrecorded site that may prove to be Site Q (for "que?"["which?"]). The potential existence of this Mayan site has been based on relief sculptures found in various collections and museums, as first identified by Peter Mathews in the late 1970s. The location of Site Q, signified by a snake-head emblem glyph, has been a source of debate among scholars, with some postulating that the site has not yet been found and others claiming that it is the known site Calakmul in Mexico (Schuster, 1997). The preeminence of Site Q in the Classic Maya world is evidenced by the fact that the snake-head emblem glyph has been found in inscriptions at sites in Belize, Guatemala, Honduras, and Mexico. While Calakmul is a strong candidate for Site Q, the fact remains that there are a number of large Mayan sites in the area where little or no research has been conducted, not to mention potential sites that have not yet been found.

While studying scarlet macaw routes, team member Santiago Billy came across an unrecorded site in April 1996. His GPS measurements of this location were compared with TM satellite imagery in which potential causeway lineaments could be seen by the author. Our team conducted an initial reconnaissance mission in July 1996. Pyramids, walls, plazas, structures, and hieroglyphic monuments could be seen. The research team reported this information to Ian Graham and David Stuart at the Harvard Peabody Museum.

In May 1997, our research team met Graham and Stuart at the site, which has been named "La Carona." As they mapped the site and recorded the hieroglyphs, we surveyed the site perimeter, finding many additional but smaller structures. To date, sufficient analysis has been conducted to determine that if this location is not Site Q itself, it is nearby (Graham, 1997). Our remote sensing imagery suggests that there are other archeological sites in the area that we intend to visit in the course of our research. (The difficulties of conducting ground-truth reconnaissance in these areas are discussed later in this chapter.)

DEFORESTATION

Central America has one of the highest rates of deforestation, on a percentage basis, in the world (Food and Agriculture Organization, 1991). Satellite imagery provides the primary source of quantifiable data about forest and land-cover changes. The remote sensing and GIS techniques used for deforestation and vegetation mapping are well established (Fearnside, 1986; Sader and Joyce, 1988; Sader et al., 1990; Skole and Tucker, 1993). These analytical techniques have been applied to present evidence of the current deforestation threat to the Peten region.

As noted earlier, the ancient Maya had substantially deforested the region by the time of their collapse in the ninth century, while the current tropical forest, no more than 1,000 years old, is being destroyed at an alarming rate in the wake of human migration and settlement. Schwartz (1990) documents the social and political driving forces that have affected the people and the forest of the region. Until recently, the Peten enjoyed a largely stable sociocultural system dating back to the 1720s. From colonial times it had been a sparsely populated region, with one writer commenting, "After all, the Peten was then (1895) and is now (1940) a region almost unoccupied, little developed, and of small promise" (Jones, 1940, quoted in Schwartz [1990:241]). The region remained sparsely populated until the 1960s, when the national government opened the Peten to colonization. Unequal land distribution, decreasing access to land, extensive deforestation, increasing economic inequality, and political unrest subsequently led to revolution.

By the late 1970s and early 1980s revolutionary guerrilla forces had established a wide base of insurgency support against the national government in both the Peten and the Guatemala highlands. The conflict had reached a peak by the mid-1980s, but the struggle continued until the signing of a peace treaty between the rebels and the government in January 1997. Between 1960 and 1986, the Peten experienced a population increase from 26,000 to over 300,000—an increase of 1,100 percent in 25 years (Schwartz, 1990:256-257). As will be seen, our satellite imagery demonstrates that between 1986 and 1995, human migration and deforestation continued to accelerate.

Tropical rainforest deforestation results from complex combinations of social, economic, and biological causes. It affects the flora and fauna and the lifestyles of indigenous populations (Nations, 1988). Currently, cattle ranching and slash-and-burn agriculture are the most serious threats to the Peten. Others include road building, oil exploration, selective logging of mahogany, the taking of endangered species of animals and birds, and the looting of archeological treasures. Since the late 1970s, marijuana has also become an important though illegal export crop (Schwartz, 1990:260). These practices stand in contrast to economic activities that have a negligible environmental impact, such as tourism and the harvesting of chicle, xate, and pimienta.

Four Landsat TM satellite scenes were purchased for April 1986, April 1990, May 1993, and March 1995. As with many tropical regions, it is difficult to acquire cloud-free data over the Peten, and these satellite scenes represent the best data available. Image processing was conducted using Earth Resources Laboratory Applications Software (ELAS) (Graham et al., 1985) and ERDAS (1990) software. Data processing and analysis were conducted jointly under the direction of Steven Sader, University of Maine, and the author. A channel 3, 4, and 5 subset was extracted from the seven-channel TM data. Atmospheric haze was reduced by subtracting the minimum values located in deep water (Lago Peten Itza) from the visible red (channel 3) and reflective near infrared (channel 4). The NDVI was computed for all four dates. The equation for NDVI is:

$$NDVI = (near\ infrared - visible\ red)/(near\ infrared + visible\ red)$$

NDVI information correlates with measures of vegetation "greenness" (Tucker, 1979; Sellers, 1985) and can easily distinguish roads, water, and sparse vegetation from high biomass forests (Sader et al., 1991). Change detection was performed by subtracting the 1990, 1993, and 1995 NDVI from the 1986 NDVI. The data were compared with false color composite images. The composites were made by loading TM bands 4, 5, and 3 onto the red, green, and blue image planes of the computer screen (Jensen, 1986). In this way a time-sequence of change from 1986 can be identified and quantified. The results are alarming for the MBR, an area generally thought of as being protected.

Increased forest clearing can be identified throughout various parts of the MBR. Comparison of the satellite images for the above four dates indicates that the greatest deforestation of primary forest occurred during 1990-1993, although the Belize-Guatemala border area saw the greatest increase during 1993-1995. Interior roadless regions within the MBR show little or no regrowth or clearing activity. There is a strong relationship between roads and distance to forest, with over 90 percent of the clear cutting being conducted within 2 km of roads. What is certain is that the ratio of clear-cutting to mature forest has been increasing since 1986. Results indicate a general 12:1 ratio (12 ha cut to 1 ha regrown) during 1986-1990. During 1990-1993, some areas experienced even higher ratios. We are currently completing our analysis for 1993-1995, but the same pattern of increasing deforestation is apparent as the forested landscape with scattered agricultural fields yields to an agricultural landscape with a fragmented forest.

Forest-clearing ratios appear to be highest in frontier areas where there are new roads and an influx of settlers. These areas include the Sierra del Lacandon and the Mexico-Guatemala border region in the west. Forest clearing is less severe near lower, more stable resident populations, as in Carmelita in the central Peten, where more traditional farming techniques are practiced. Forest clearing is also less severe in areas where there are guards on duty, such as Tikal.

It is not simply the ratios of deforestation that are alarming, but also the pattern of deforestation. Forest-clearing activities have created "vegetation islands" of primary forest surrounded by cleared areas. These vegetation islands are becoming more pronounced in response to new settlers and the construction of new roads and an oil pipeline in the area (apparent in the 1995 TM image). The extent to which this increasing deforestation is affecting the ecology of the region, in particular the migration routes of birds and animals, is unknown at this time.

What is known is that the increasing human migration into the area is adding to the social stress as the competition for land and resources increases. Satellite change-detection images can be used to identify potential areas of conflict as the forces of expansion meet protected land. The peace treaty of 1997 provides for the resettlement within the MBR of refugees who fled the area during the conflict. The resulting impact on the forest will be seen in future satellite images. Between March and May 1997, there were two serious incidents inside the MBR, in which government and nongovernmental organization (NGO) hostages were captured at gunpoint, and a biological station was burned to the ground by a force of 60 armed men. Eventually, the release of the hostages was negotiated, and an agreement was made with the perpetrators to rebuild the biological station in exchange for logging, farming, and hunting concessions within the reserve. In May 1997, Carlos Catalan, a Guatemalan spokesman for conservation, was murdered at gunpoint in the village of Carmelita. In June 1997, archeologist Peter Mathews and his five-man research team were captured and beaten along the Usumacinta River as they attempted to remove a Mayan altar to a local museum. Although Matthews had the necessary government and local permissions, the captives told him that "they don't respect any authority" (Associated Press, 1997). These recent incidents illustrate the reality of the situation and suggest the likelihood that similar incidents will occur in the future unless an effective land-tenure and migration policy can be established and enforced.

RESEARCH CHALLENGES:
COLLECTING GROUND-TRUTH INFORMATION

The gathering of ground-truth information, often referred to as "reference data," involves collecting measurements or observations about objects, areas, or phenomena that are being remotely sensed (Lillesand and Kiefer, 1994). Social scientists can use ground-truth information in two ways: first, it can aid in the analysis, interpretation, and validation of the remotely sensed data; second, such information helps in understanding the socioeconomic forces behind land-cover modifications due to human activities.

Ground truthing is expensive and time-consuming. While the price of computer hardware and software for remote sensing analysis has dropped dramatically in recent years, the costs associated with ground-truth activities have generally increased. Airfare, lodging, vehicle rental, food, labor, and the like remain expensive elements of a research design, although it may be noted that recent advances in affordable, portable GPS receivers and digital field recorders give the field investigator greater flexibility in the sampling design.

Our ground-truth activities require that we visit as many sites as possible in the remote and difficult terrain of the Peten in order to document as many study sites as possible. Consequently, we tend to break camp each day and move on to the next location. This feature of our research contrasts with the approach of those who remain at a village or site for an extended period of time. While we map and verify the existence of archeological features, we never excavate. Some of the challenges we have encountered while conducting fieldwork in the Peten include logistical problems, communication problems, equipment failure, inadequate maps, physical stress, suspicion, danger, and unstable political environments. In fact, our research team was captured and held at gunpoint by leftist guerrillas for several hours before being released at nightfall.

Certain phenomena (such as prehistoric Mayan roadways) that appear in the nonvisible bandwidths of the remote sensing data are sometimes simply not visible from ground level and require clearing and/or excavation before being verified. This situation is related to the fact that certain features that cannot be seen in visible light can nevertheless be detected in the infrared or microwave portions of the spectrum by remote sensing instrumentation (Sever, 1990). Field reconnaissance also provides us with other information not visible in the imagery, such as the selective cutting of mahogany trees. Generally, our interviews with local farmers and ranchers provide accurate information. Sometimes, however, the respondents provide us with either information they think we want to hear or false information intended to mask illegal activity.

Logistics are probably the major constraint on our field work. Often we are the first professionals to visit an unrecorded archeological site. We must schedule in advance the jeeps, boats, aircraft, mules, horses, and workers that will get us to our destination. Since many areas of the Peten do not have telephone service, a member of our team who lives in Guatemala must travel weeks and months in advance to arrange these rentals with the local villagers. The more inaccessible the location, the more difficult the arrangements are. Generally, our field missions last 2 to 3 weeks. As we travel and switch from jeeps to boats to horses and mules, it is critical that the dates, times, and locations for these arrangements be finalized in advance.

Occasionally, we are met with suspicion regarding the true purpose of our research. We have successfully combatted this situation by taking the time to explain our research goals and objectives to the local residents. We always take a large number of satellite images and, after explaining how we are using the imagery, leave a copy of it with the person(s) involved. Through years of exposure and word of mouth, we have gained acceptance and support for our data acquisition activities, as well as the confidence of the inhabitants. As the years have passed, many of the inhabitants have become better educated about satellite imagery and GPS units, so that when we stop in a village and present the images, they can often help us interpret some of the features and anomalies displayed. Having a Guatemalan national on the research team is also a positive benefit to our research activities.

GPS measurements are a critical component of our field research. Initially, in 1988, there were only a few satellites in orbit, and we often found ourselves climbing a temple at midnight to collect a position that would be available only between 1:00 and 4:00 a.m. Also, intense vegetation cover of the tropical forest limited the collection of GPS readings. Today there is a complete constellation of satellites, and readings can be gathered nearly around the clock. New GPS units with better software allow for the collection of more data with greater accuracy.

A problem we often encounter is the inaccuracy of available maps. We consistently find that lakes, rivers, archeological sites, and cultural features are not located where the map indicates. Apart from the obvious in-field confusion, there is the major concern that if these inaccurate maps are digitized and incorporated into a GIS, they will lead to false results for predictive models. We resolve this dilemma by constantly comparing our GPS measurements, imagery, and maps to eliminate as much confusion as possible. It should also be noted that the names on a map are not necessarily the names used by the local inhabitants.

As we studied deforestation trends in the Peten over the last several years, we designed our ground-truth activities primarily to identify the difference between new forest clearings and regrowth. Now we are expanding our ground-truth activities to include information on the decision processes associated with land use and land conversion. Currently, the various socioeconomic factors associated with deforestation are poorly understood. In addition, much of the current uncertainty involved in modeling the terrestrial carbon budget arises from inadequate data on tropical deforestation rates and trends in land-use conversion. To address these scientific issues, we will acquire additional information as we interview local farmers and ranchers. Through these interviews, we will determine such factors as the crop-to-fallow ratios, the decision process for converting land to pasture or to shifting agriculture, the forest fragmentation indices and spatial characteristics of cleared land, the associated socioeconomic factors, and how the driving forces differ by zones or management units. The results will be correlated over the time scale of our database, providing better analytical information for management decisions.

SUMMARY AND CONCLUSIONS

The delicate ecosystem of the MBR in the Peten region of northern Guatemala was managed successfully for centuries by the ancient Maya until their collapse in the ninth century A.D. The archeological evidence indicates, however, that by the time of their collapse, the Maya had deforested most of the region. For the next 1,000 years the forest regenerated. Today, human migration and so-called "modern" subsistence techniques once again threaten the sustainability of the area.

Through the use of remote sensing and GIS research, we are attempting to answer questions about the past in order to protect and manage the resources of the future. The protection of the tropical rainforest is also synonymous with the protection of archeological sites, for the cutting and burning of an area destroy not only the trees, but also archeological features and materials. In addition, the construction of modern roads and pipelines provides a conduit for the illegal looting of endangered animal species and archeological treasures.

Remote sensing data have successfully located archeological features in the Peten region that are difficult to discern from traditional survey and analysis techniques. These features include ancient roadways, temples, cisterns, and agricultural areas. The detection and subsequent analysis of these features may help answer questions regarding ancient Mayan subsistence, transportation, and water management and the factors that led to the eventual abandonment of the area at the time of the Mayan collapse. At the same time, however, verification of archeological features detected in the remote sensing data is expensive, time-consuming, difficult, and dangerous because of the dense vegetation, remoteness, and hazards associated with the Peten landscape.

After the ninth century, the Peten remained a sparsely populated area until the Guatemalan government opened the area to colonization in the 1960s. Since that time the forest has continued to be cut down at an accelerating rate as human migration and settlement have introduced nontraditional agricultural techniques. The driving forces behind deforestation activities in the region result from a complex combination of social, economic, and political issues. Change-detection analysis using remote sensing imagery has been used to document and quantify the extent and rate of change during 1986-1995. Data analysis reveals that the ratio of clear-cutting to mature forest has been increasing over that time. The highest forest-cutting ratios are associated with the construction of new roads and increases in migrating settlers. Forest-clearing ratios are lowest near resident populations who practice traditional farming techniques. Cattle ranching and slash-and-burn agriculture are the greatest threats to the Peten, although road building, oil exploration, selective logging of mahogany, and illicit crops also adversely impact the ecology of the region.

Increasing human migration and settlement in the next few years will only add to the deforestation and social stress as the competition for resources increases. According to current estimates, only 2 percent of the Peten's forest will survive by the year 2010. The use of aerial and satellite imagery can help identify areas of possible conflict, inventory natural and cultural resources, identify illegal practices, and monitor protected areas. It is hoped that the results of this interdisciplinary research will provide information that will help managers, scientists, and politicians responsible for the MBR and the Peten region in general make more informed decisions and thereby avoid the collapse that occurred in this area a little over 1,000 years ago.

REFERENCES

Adams, R.E.W., T.P. Culbert, W.E. Brown, P.D. Harrison, and L.J Levi
  1990   Rebuttal to Pope and Dahlin. *Journal of Field Archaeology* 17:241-243.
Associated Press
  1997   Archeologists barely survive ambush in Mexican jungle. *CNN Interactive World News,* July 1.
Cahuec, E., and J. Richards
  1994   La variacion sociolinguistica del Maya Q'eqchi. *Boletin Linguistico* 40 Guatemala.
Canteo, C.
  1996   Destruccion de Biosfera Maya avanza ano con ano. *Siglo Veintiuno* (Guatemala), November 21.
Culbert, T.P.
  1993   *Maya Civilization.* Washington, D.C.: St. Remy Press and Smithsonian Institution.
  1974   *The Lost Civilization: The Story of the Classic Maya.* New York: Harper and Row.
Culbert, T.P., L. Levi, B. McKee, and J. Kunen
  1995   Investigaciones Arqueologicas en el Bajo La Justa, Entre Yaxha y Nakun. Unpublished field report, Department of Anthropology, University of Arizona.
Curtis, J.H., D.A. Hodell, and M. Brenner
  1996   Climate variability on the Yucatan Peninsula (Mexico) during the past 3500 years, and implications for Maya cultural evolution. *Quaternary Research* 46, Article No. 0042:37-47.
ERDAS
  1990   Field Guide, Version 7.4. ERDAS, Inc., Atlanta, Ga.
Fearnside, P.M.
  1986   Spatial concentration of deforestation in the Brazilian Amazon. *Ambio* 15:74-81.
Food and Agriculture Organization
  1991   Second Interim Report on the State of Tropical Forests. Unpublished paper, Forest Resources Assessment 1990 Project, United Nations FAO, Rome, Italy.
Graham, I.
  1997   Mission to La Carona. *Archaeology,* September/October:46.
Graham, M.H., B.G. Junkin, M.T. Kalcic, R.W. Pearson, and B.R. Seyfarth
  1985   Earth Resources Laboratory Applications Software (ELAS), Revision 3. NASA-NSTL-Earth Resources Laboratory, Report No. 183, Stennis Space Center, Miss.
Hammond, N.
  1990   *Ancient Maya Civilization.* New Brunswick, N.J.: Rutgers University Press.
Jensen, J.R.
  1986   *Introductory Image Processing—A Remote Sensing Perspective.* Englewood Cliffs, N.J.: Prentice-Hall.
Jones, C.L.
  1940   *Guatemala: Past and Present.* Minneapolis: University of Minnesota Press.
Kelley, P.S.
  1983   Digital analysis in remote sensing: from data to information. Pp. 218-240 In *Remote Sensing of the Environment,* B.F. Richason, Jr. ed.
Lillesand, T.M., and R. Kiefer
  1994   *Remote Sensing and Image Interpretation.* New York: John Wiley and Sons.
Lundell, C.L.
  1937   The Vegetation of the Peten. Carnegie Institute Publication No. 478, Washington, D.C.
Nations, J.
  1988   *Tropical Rainforest: Endangered Environment.* New York: Franklin Watts.

Pope, K.D., and B.H. Dahlin
   1989   Ancient Maya wetland agriculture: New insights from ecological and remote sensing. *Journal of Field Archaeology 16:87-106.*
   1993   Radar detection and ecology of ancient Maya canal systems. *Journal of Field Archaeology* 20:379-383.
Rice, D.S.
   1991   Roots. *Natural History* February:10-14.
Sabloff, J.A.
   1995   Drought and decline. *Nature* 375: 357.
Sader, S.A.
   1995   Spatial characteristics of forest clearing and vegetation regrowth as detected by Landsat Thematic Mapper imagery. *Photogrammetric Engineering and Remote Sensing* 61:1145-1151.
Sader. S.A., and A.T. Joyce
   1988   Deforestation rates and trends in Costa Rica, 1940-1983. *Biotropica* 20(1):11-19.
Sader, S.A., T.A. Stone, and A.T. Joyce
   1990   Remote sensing of tropical forests: an overview of research and applications using non-photographic sensors. *Photogrammetric Engineering and Remote Sensing* 55(10):1343-1351.
Sader, S.A., G.V.N. Powell, and J.H. Rappole
   1991   Migratory bird habitat monitoring through remote sensing. *International Journal of Remote Sensing* 12(3):363-372.
Sader, S.A., T. Sever, J.C. Smoot, and M. Richards
   1994   Forest change estimates for the northern Peten region of Guatemala—1986 to 1990. *Human Ecology* 22(3):317-322.
Sader, S.A., T. Sever, and J.C. Smoot
   1996   Time series tropical forest change detection: A visual and quantitative approach. International Symposium on Optical Science, Engineering and Instrumentation, Denver, Colo. *SPIE*, Volume 2818-2-12.
Schuster, A.M.H.
   1997   The search for Site Q. *Archaeology* September/October:42-45.
Schwartz, N.B.
   1990   *Forest Society: A Social History of Peten, Guatemala.* Philadelphia: University of Pennsylvania Press.
Sellers, P.J.
   1985   Canopy reflectance, photosynthesis and transpiration. *International Journal of Remote Sensing* 6(8):1335-1372.
Sever, T.
   1990   Remote Sensing Applications in Archeological Research: Tracing Prehistoric Human Impact Upon the Environment. Ph.D. dissertation, University of Colorado, Boulder. Available: University Microfilms, Ann Arbor, MI.
Sheets, P., and T. Sever
   1988   High tech wizardry. *Archaeology* 41(6) Nov/Dec:28-35.
Skole, D., and C. Tucker
   1993   Tropical deforestation and habitat fragmentation in the Amazon: Satellite data from 1978 to 1988. *Science* 260:1905-1910.
Stuart, G.E.
   1991   Maya heartland under siege. *National Geographic* 10:94-107.
Trombold, C.D.
   1991   *Ancient Road Networks and Settlement Hierarchies in the New World.* Cambridge, England: Cambridge University Press.

Tucker, C.J.
  1979   Red and photographic infrared linear combinations for monitoring vegetation. *Remote Sensing Environment* 8:127-150.
Turner II, B.L.
  1993   Rethinking the "new orthodoxy:" Interpreting ancient Maya agriculture. Pp. 57-88 In *Culture, Form, and Place: Essays in Cultural and Historical Geography,* K. Mathewson, ed. Geoscience and Man 32. Baton Rouge, La.: Louisiana State University.

8

Extraction and Modeling of Urban Attributes Using Remote Sensing Technology

David J. Cowen and John R. Jensen

Since the earliest developments in urban sociology and geography (Harris and Ullman, 1945), researchers have recognized the essential spatial element of urban development. Remote sensing provides an opportunity to measure attributes of urban and suburban environments and record the data in accurate digital maps and files suitable for analysis with geographic information systems (GIS). These data, together with data available from ground-based observations, can be used to monitor changes in space and time, to develop and validate dynamic models of urban development, and to forecast future land-use patterns and changes in other urban attributes. Remotely sensed data are thus potentially valuable both to social scientists and to urban planners and other public officials.

This chapter first identifies key attributes of urban and suburban environments and evaluates the capability of remote sensing technology to measure these attributes accurately at the requisite levels of temporal, spatial, and spectral resolution. It then presents a detailed case example that illustrates how measurements of several of these attributes can be combined to address a social science problem: the development of an empirically based theory of urban residential expansion.

REMOTE SENSING OF URBAN/SUBURBAN ATTRIBUTES

Humans create complex urban landscapes that are composed of various materials (concrete, asphalt, plastic, shingles, water, grass, soil, and shrubbery) arranged in specific ways to build transportation systems, utility lines, housing, commercial buildings, and public space in order to improve the quality of life.

Characteristics of many of these phenomena can be remotely sensed from subor-bital aircraft or from satellites. The information thus derived may be both quali-tative and quantitative.

Ten of the major urban/suburban attributes of significant value for under-standing the urban environment are summarized in Table 8-1. To remotely sense these urban phenomena, it is necessary to understand the temporal and spatial resolution required for each. Temporal resolution refers to how often managers need the information; for example, local planning agencies may need precise population estimates every 5 to 7 years to supplement estimates provided by the decennial census. As an example of required spatial resolution, local population estimates based on building unit counts usually must have a minimum mapping unit of 0.3-5 m (0.98-16.4 ft). The information presented in Table 8-1 was synthesized both from the literature (e.g., Branch, 1971; Ford, 1979; Jensen, 1983a, b; Haack et al., 1997; Philipson, 1997) and from practical experience. Ideally, there would always be a remote sensing system that could obtain images of the terrain that would satisfy the temporal and spatial resolution requirements specified in Table 8-1. Unfortunately, this is not always the case, as will be demonstrated later in this chapter.

Information about urban attributes is also best collected using the specific portions of the electromagnetic spectrum shown in Table 8-1. For example, land cover (U.S. Geological Survey [USGS] Level III) is best acquired using the visible (V: 0.4-0.7 micrometers [μm]), near-infrared (NIR: 0.7-1.1 μm),,and mid-infrared (MIR: 1.5-2.5 μm) portions of the spectrum. Building perimeter, area, volume, and height information is best acquired using black-and-white pan-chromatic (0.5-0.7 μm) or color imagery. The thermal infrared portion of the spectrum (TIR: 3-12 μm) may be used to obtain urban temperature measurements.

The relationship between temporal and spatial data requirements for urban/suburban attributes and the temporal and spatial characteristics of available and proposed remote sensing systems is shown in Figure 8-1. Note that the codes shown on this figure are defined in Table 8-1, while abbreviations used for the various remote sensing systems are defined in the glossary in Appendix B, and in Morain and Budge (1996).

Land Use/Land Cover

Urban land-use/land-cover information is required for residential-industrial-commercial site selection, population estimation, and development of zoning regulations (Green et al., 1994). For this reason, the USGS developed a land-use and land-cover classification system for use with remotely sensed data (Anderson et al., 1976). Broad Level I classes may be inventoried using the Landsat Multi-spectral Scanner (MSS), with a spatial resolution of 79 × 79 m; the Thematic Mapper (TM), with a resolution of 30 × 30 m; the Système pour l'Observation de la Terre (SPOT) High Resolution Visible (HRV) (XS), with a resolution of 20 ×

TABLE 8-1 Relationship Between Selected Urban/Suburban Attributes and the Remote Sensing Resolutions Required to Provide Such Information

	Minimum Resolution Requirements		
	Temporal	Spatial	Spectral
Land Use/Land Cover			
L1 - USGS Level I	5 -10 years	20 - 100 m	V-NIR-MIR-Radar
L2 - USGS Level II	5 -10 years	5 - 20 m	V-NIR-MIR-Radar
L3 - USGS Level III	3 - 5 years	1 - 5 m	V-NIR-MIR-Pan
L4 - USGS Level IV	1 - 3 years	0.3 - 1 m	Pan
Building and Property Line Infrastructure			
B1 - building perimeter, area, volume, height	1 - 2 years	0.3 - 0.5 m	Pan
B2 - cadastral mapping (property lines)	1 - 6 months	0.3 - 0.5 m	Pan
Transportation Infrastructure			
T1 - general road centerline	1 - 5 years	1 - 30 m	Pan
T2 - precise road width	1 - 2 years	0.3 - 0.5 m	Pan
T3 - traffic count studies (cars, airplanes etc.)	5 - 10 min	0.3 - 0.5 m	Pan
T4 - parking studies	10 - 60 min	0.3 - 0.5 m	Pan
Utility Infrastructure			
U1 - general utility line mapping and routing	1 - 5 years	1 - 30 m	Pan
U2 - precise utility line width, right-of-way	1 - 2 years	0.3 - 0.6 m	Pan
U3 - location of poles, manholes, substations	1 - 2 years	0.3 - 0.6 m	Pan
Digital Elevation Model (DEM) Creation			
D1 - large scale DEM	5 - 10 years	0.3 - 0.5 m	Pan
D2 - large scale slope map	5 - 10 years	0.3 - 0.5 m	Pan
Socioeconomic Characteristics			
S1 - local population estimation	5 - 7 years	0.3 - 5 m	Pan
S2 - regional/national population estimation	5 - 15 years	5 - 20 m	V-NIR
S3 - quality of life indicators	5 - 10 years	0.3 - 30 m	Pan-NIR
Energy Demand and Conservation			
E1 - energy demand and production potential	1 - 5 years	0.3 - 1 m	Pan-NIR
E2 - building insulation surveys	1 - 5 years	1 - 5 m	TIR
Meterological Data			
M1 - daily weather prediction	30 min - 12 hr	1 - 8 km	V-NIR-TIR
M2 - current temperature	30 min - 1 hr	1 - 8 km	TIR
M3 - current precipitation	10 - 30 min	4 km	Doppler Radar
M4 - immediate severe storm warning	5 - 10 min	4 km	Doppler Radar
M5 - monitoring urban heat island effect	12 - 24 hr	5 - 10 m	TIR
Critical Environmental Area Assessment			
C1 - stable sensitive environments	1 - 2 years	1 - 10 m	V-NIR-MIR
C2 - dynamic sensitive environments	1 - 6 months	0.3 - 2 m	V-NIR-MIR-TIR
Disaster Emergency Response			
DE1 - pre-emergency imagery	1 - 5 years	1 - 5 m	V-NIR
DE2 - post-emergency imagery	12 hr - 2 days	0.3 - 2 m	Pan-NIR-Radar
DE3 - damaged housing stock	1 - 2 days	0.3 - 1 m	Pan-NIR
DE4 - damaged transportation	1 - 2 days	0.3 - 1 m	Pan-NIR
DE5 - damaged utilities	1 - 2 days	0.3 - 1 m	Pan-NIR

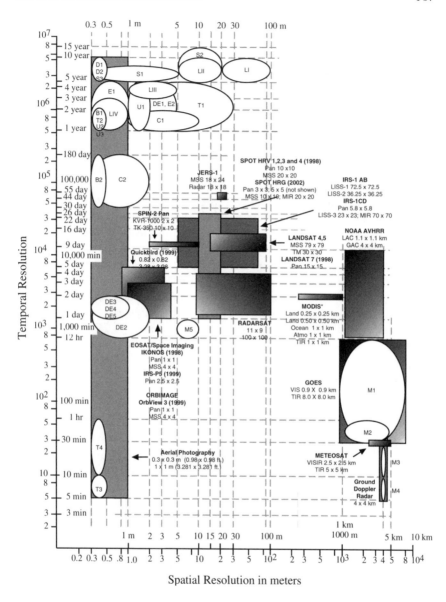

FIGURE 8-1 Spatial and temporal resolution requirements for urban/suburban attributes overlaid on the spatial and temporal capabilities of current and proposed remote sensing systems.

20 m; and the Indian LISS 1-3 (72 × 72, 36.25 × 36.25, and 23.5 × 23.5 m, respectively). For example, Plate 8-1 (after page 182) depicts USGS Level I urban vs. nonurban information for Charleston, South Carolina, extracted from Landsat data for 1973, 1981, 1982, and 1994. Sensors with a minimum spatial resolution of 5-15 m (e.g., SPOT panchromatic [pan] at 10 × 10 m; SPIN-2 TK-350 at 10 × 10 m; proposed Landsat 7 pan at 15 × 15 m) are required to obtain USGS Level II information, which includes specific types of man-made structures. USGS Level III classes may be inventoried using sensors with a spatial resolution of 1-5 m, such as Indian Remote Sensing (IRS) pan (approximately 5 × 5 m) and the SPIN-2 KVR-1000 (2 × 2 m). Future sensors may include commercial ventures such as EOSAT/Space Imaging IKONOS (1 × 1 m pan), OrbView 3 (1 × 1 m pan), and Indian IRS P5 (2.5 × 2.5 m) (Montesano, 1997). USGS Level IV classes may best be monitored using high-spatial-resolution sensors, including aerial photography (0.3-1 m), and proposed EarthWatch Quickbird pan (0.8 × 0.8 m) and IKONOS (1 × 1 m). A sensor that collects panchromatic data of 0.3-0.5 m resolution is required to provide detailed Level IV information. RADARSAT provides data with 11-100 m spatial resolution for Level I and II land-cover inventories, even in cloud-shrouded tropical landscapes where conventional sensors would not be able to penetrate (Leberl, 1990).

Urban land-use/land-cover classes in Levels I through IV have temporal resolution requirements of 1-10 years (see Table 8-1 and Figure 8-1). All of the sensors mentioned have temporal resolutions of less than 22 days, and thus satisfy these requirements.

Building and Cadastral (Property Line) Infrastructure

Data on building perimeter, area, volume, and height are best obtained using stereoscopic (overlapping) panchromatic aerial photography or other remote sensing data with a spatial resolution of 0.3-0.5 m (Jensen, 1995; Warner et al., 1996). The stereo images are required to visualize features in three dimensions. For example, panchromatic stereoscopic aerial photography with a spatial resolution of 0.3 × 0.3 m (1 ft) was used to extract building perimeter and area information for a residential area in Covina, California (Figure 8-2). Each building, tree, driveway, fence, and contour can be extracted from this type of data. In many instances, the fence lines are the cadastral property lines. Accurate photogrammetric surveys can meet the new draft Geospatial Positioning Accuracy Standards (Federal Geographic Data Committee, 1997). If necessary, the property lines can be surveyed by a licensed surveyor and the information overlaid onto the photographic or planimetric map database to represent the legal cadastral (property) map. Many municipalities in the United States are moving toward using such high-spatial-resolution imagery as the source for some cadastral information and as an image backdrop upon which to depict all surveyed cadastral information.

Detailed data on building height and volume can be extracted from high-spatial-resolution (0.3-0.5 m) stereoscopic imagery (Jensen et al., 1996). Such information can then be used to create three-dimensional displays of the terrain that one can walk through in a virtual-reality environment if desired (Wolff and Yaeger, 1993) (see Figure 8-3). Such information provides an extremely useful way to visualize the density and arrangement of structures in a neighborhood, and architects, planners, engineers, and realtors are beginning to use this information for a variety of purposes. It is expected that in the next few years, Space Imaging (1997) and EarthWatch (Quickbird, 1998/1999) will provide such stereoscopic images from satellite-based platforms with approximately 0.8-1 m spatial resolution. Unfortunately, such imagery will still not provide the detailed planimetric (perimeter, area) and topographic (terrain contours, building height and volume) details that can be extracted from high-spatial-resolution large-scale aerial photography. Therefore, a satellite sensor system with 0.3-0.5 m spatial resolution may be required, but it will not be available in the immediate future.

Transportation Infrastructure

Transportation studies have long relied on remote sensor data to (1) examine the origin and destination of trips; (2) study traffic patterns at choke points such as tunnels, bridges, shopping malls, and airports; (3) analyze metropolitan traffic patterns; (4) conduct parking studies; and (5) evaluate the condition of roads (Mintzer, 1983; Haack et al., 1997). The general updating of a road network centerline map is a fundamental task that is often done once every 1 to 5 years. In areas with minimum tree density, this task can be accomplished using imagery with a spatial resolution of 1 to 30 m (Lacy, 1992). If more precise road dimensions, such as the exact width of the road and sidewalks, are needed, a spatial resolution of 0.3-0.5 m is required (Jensen et al., 1994). Currently, only aerial photography can provide such planimetric information (see Figure 8-2).

Next to meteorological investigations, traffic-count studies of automobiles, airplanes, boats, pedestrians, and people in groups require data of the highest temporal resolution, ranging from 5 to 10 minutes. It is difficult to resolve the type of car or boat using even 1 × 1 m data; for this purpose, high-spatial-resolution imagery (0.3-0.5 m) is required. Such information can be acquired only via aerial photography or video sensors that are (1) located on the top edges of buildings looking obliquely at the terrain, or (2) placed in aircraft or helicopters and flown repeatedly over the study areas. Parking studies require the same high spatial resolution (0.3-0.5 m) but slightly lower temporal resolution (10-60 minutes). Road and bridge conditions (e.g., cracks, potholes) can be documented using high-spatial-resolution aerial photography (<0.3 × 0.3 m) (Stoeckeler, 1979).

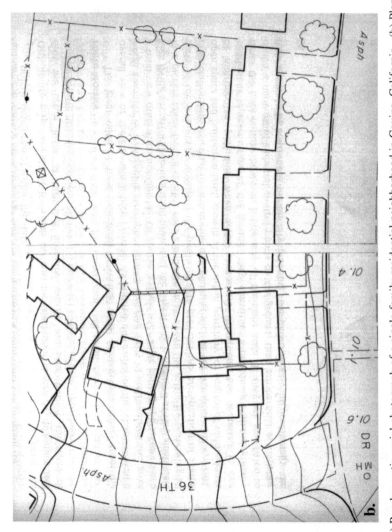

FIGURE 8-2 (a) Panchromatic aerial photograph of a single-family residential neighborhood in Covina, California. (b) Planimetric map of the area derived from a stereomodel, including building perimeter, fence lines, driveways, major trees, street curbs, and telephone poles. Topographic 2-ft contours are also provided, including spot elevations.

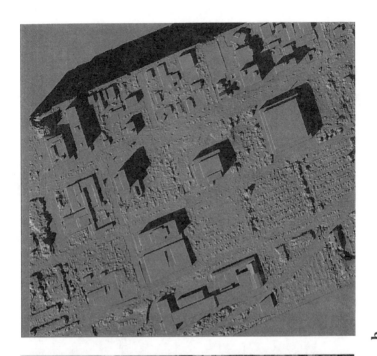

b

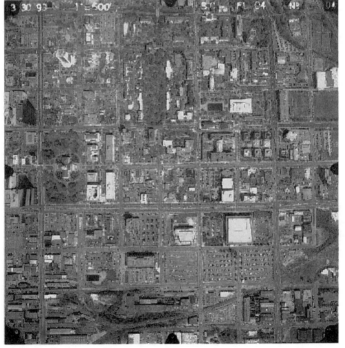

a

FIGURE 8-3 (a) Panchromatic photograph of Columbia, South Carolina (original at 1:6,000 scale). (b) Digital elevation model (DEM) derived from a stereopair and portrayed as a shaded-relief model. (c) Orthophotograph of the area draped over the DEM and displayed in a three-dimensional perspective projection. (d) Identification of the optimum building on which to place a cellular phone transceiver in downtown Columbia, South Carolina, using the DEM and a viewshed model.

Utility Infrastructure

Urban/suburban environments are enormous consumers of electrical power, natural gas, telephone service, and potable water (Haack, 1997). In addition, they create great quantities of garbage, waste water, and sewage. The removal of storm water from urban impervious surfaces is also a serious problem (Schultz, 1988). Automated mapping/facilities management (AM/FM) and GIS have been developed to manage extensive right-of-way corridors for various utilities, especially pipelines (Jadkowski et al., 1994). The most fundamental task is updating maps to show a general centerline of the utility of interest, such as a powerline right-of-way. This task is relatively straightforward if the utility is not buried. Major utility rights-of-way can be observed well in imagery with a spatial resolution of 1-2 m and obtained once every 1-5 years. However, when it is necessary to inventory the exact locations of footpads or transmission towers, utility poles, manhole covers, the true centerline of the utility, the width of the utility right-of-way, and the dimensions of buildings, pumphouses, and substations, a spatial resolution of 0.3-0.6 m is required (Jadkowski et al., 1994).

Creation of a Digital Elevation Model

Almost all GISs used for socioeconomic or environmental planning include a digital elevation model (DEM) (Cowen et al., 1995). Analysts often forget that DEMs are derived from analysis of stereoscopic remote sensor data (Jensen, 1995). It is possible to extract z-elevation data using SPOT 10×10 m data and even Landsat TM 30×30 m data for terrain that has not been mapped previously (Gugan and Dowman, 1988). However, any DEM to be used for an urban/ suburban application should ideally have a z-elevation and x, y coordinates that meet Draft Geospatial Positioning Accuracy Standards (Federal Geographic Data Committee, 1997). At a minimum, the data should meet the old USGS national map accuracy standards. The only sensor that can provide such information at the present time is stereoscopic large-scale metric aerial photography with a spatial resolution of 0.3-0.5 m. The terrain elevation does not change very rapidly. Therefore, a DEM of an urbanized area need be acquired only once every 5 to 10 years unless there is significant development, and the analyst wishes to compare two DEMs for different dates to determine changes in terrain elevation and identify unpermitted additions onto buildings or changes in building heights. DEM data can be modeled to compute slope and aspect statistical surfaces for a variety of applications (Jensen, 1996). They can also be used to predict the optimum sites for locating various utilities, as shown earlier in Figure 8-3(d). Digital desktop soft-copy photogrammetry is revolutionizing the creation and availability of special-purpose DEMs by minimizing the need for expensive specialized steroplotting equipment (Petrie and Kennie, 1990; Jensen, 1995).

Socioeconomic Characteristics

Numerous studies have documented the ability to extract socioeconomic attributes directly from remote sensor data or indirectly by means of surrogate information derived from the imagery. One of the most important of these attributes is population estimates. These estimates can be derived at local, regional, and national levels based on (1) counts of individual dwelling units, (2) measurement of land areas, and (3) land-use classification (Lo, 1995). Although remote sensing techniques provide statistical approximations that approach the values obtained in a regular census, considerable ground reference data are required to calibrate the model. It may be noted, however, that even ground-based population estimates are not very accurate (see Clayton and Estes, 1980).

The most accurate remote sensing method for computing the population of a local area is to count individual dwelling units. However, since it is not possible to determine from remotely sensed data what is occurring within a structure, these estimates require the following conditions (Lindgren, 1985; Lo, 1986, 1995; Holz, 1988; Raymondo, 1992; Haack, 1997):

- The imagery must be of sufficient spatial resolution to allow the identification of individual structures even through tree cover, and to determine whether buildings are residential, commercial, or industrial.
- Some estimates of the average number of persons per dwelling unit, such as those available from the decennial census, must be available.
- It must be assumed that all dwelling units are occupied, and only one family lives in each unit.

Such estimates are usually made every 5 to 7 years, and require high-spatial-resolution remotely sensed data (0.3-5 m). For example, individual dwelling units in a 32-census-block area of Irmo, South Carolina were recently extracted from 2.5×2.5 m aircraft multispectral data (Cowen et al., 1993). Correlation of the dwelling unit data derived from remote sensing with similar data derived from the census yielded a correlation coefficient of 0.91, which accounted for 81 percent of the variance. These findings suggest that the sensors that will be available on satellites in the next few years may have sufficient spatial resolution to provide a good source of information for monitoring the housing stock of a community on a routine basis. This capability will enable local governments to anticipate and plan for schools and other services using data with a much better temporal resolution than that offered by the decennial census. These data will also be of value for real estate, marketing, and other business applications (Lo, 1995).

The dwelling unit approach is not suitable for a regional/national census of population because it is too time-consuming and costly. Therefore, other methods have been developed for this purpose. Scientists have documented a strong relationship between the simple urbanized built-up area extracted from a re-motely sensed image and the settlement population (Tobler, 1969). Another population estimation technique is based on the use of Level I-III land-use information previously described. This approach assumes that land use in an urban area is closely correlated with population density. Researchers establish a population density for each land use on the basis of field survey or census data. Then, by measuring the total area for each land-use category, they estimate the total population for that category. Summing the estimated totals for each category provides the total population projection. Both of these methods can use data derived from multispectral remote sensors (5-20 m) every 5 to 15 years.

Studies have documented how quality-of-living indicators, such as house value, median family income, average number of rooms, average rent, education, and income can be computed by extracting the following variables from high-spatial-resolution (0.3-0.5 m) panchromatic and/or color infrared aerial photography (Monier and Green, 1953; Green, 1957; McCoy and Metivier, 1973; Tuyahov et al., 1973; Henderson and Utano, 1975; Lindgren, 1985; Haack, 1997):

- Building size (sq. ft)
- Lot size (acreage)
- Pool (sq. ft)
- Vacant lots per city block
- Frontage (sq. ft)
- Placement of house on lot (distance from street)
- Building density (%)
- Houses with driveway (%)
- Houses with garage (%)
- Number of autos visible per house
- Unpaved street (%)
- Street width (ft)
- Health of landscaping (based on near-infrared reflectance)
- Proximity to manufacturing and retail activity

These attributes derived from remote sensing must be correlated with in situ observations to compute the quality-of-living indicators. They are also sensitive to regional and even neighborhood variations. For example, the presence of swimming pools is more likely to be a good indicator of an affluent neighborhood in northern states than in the Sun Belt. Such indicators are usually collected every 5 to 10 years.

Energy Demand and Conservation

Studies have documented regional and national urban/suburban energy demand. First, the square footage of individual buildings is determined. Ground-truth information about energy consumption is then obtained for a representative sample of homes in the area, and regression relationships are derived to predict the energy consumption anticipated for a region. Similarly, it is possible to predict how much solar photovoltaic energy potential a geographic region has by modeling the individual rooftop square footage and orientation with known photo-voltaic-generation constraints. Doing so, however, requires imagery of very high spatial resolution (0.3-0.5 m) (Clayton and Estes, 1979; Angelici et al., 1980).

Numerous studies have documented how high-spatial-resolution (0.3-1 m) predawn thermal infrared imagery (3-5 μm) can be used to inventory the relative quality of housing insulation if (1) the rooftop material is known (e.g., asphalt versus wood shingles), (2) no moisture is present on the roof, and (3) the orientation and slope of the roof are known (Colcord, 1981; Eliasson, 1992). Accurate assessment of these conditions requires in situ measurements to verify the spectral signature. If energy conservation or the generation of solar photovoltaic power were important, these variables would probably be collected every 1 to 5 years.

Meteorological Data

Daily weather in urban environments affects people, schools, businesses, telecommunications, and transportation systems. Great expense has gone into the development of systems for near-real-time monitoring of frontal systems, temperature, and precipitation, and especially for severe storm warning. The public often forgets that these meteorological parameters are monitored almost exclusively by sophisticated airborne and ground-based remote sensing systems.

For example, two Geostationary Operational Environmental Satellites (GOES) are positioned at 36,000 km above the equator in geosynchronous orbits. GOES West obtains information about the western United States and is parked at 135° west longitude. GOES East obtains information about the Caribbean and the Eastern United States and is parked at 75° west longitude. Every day these systems allow millions of people to watch the progress of frontal systems that sometimes generate deadly tornadoes and hurricanes. The visible and near-infrared data are obtained at a temporal resolution of 30 minutes, with some of the images being aggregated to create 1-hour and 12-hour animation. The spatial resolution of GOES East and West is 0.9×0.9 km for the visible bands and 8×8 km for the thermal infrared band. European nations use Meteosat, with visible near-infrared data obtained at a resolution of 2.5×2.5 km and thermal infrared data collected at a resolution of 5×5 km every 30 minutes. Early hurricane monitoring and modeling based on data acquired from these systems have saved thousands of lives in recent years.

The public also relies on ground-based Doppler radar for near-real-time precipitation and severe storm warning. Doppler radar obtains data with a resolution of 4×4 km every 10 to 30 minutes when monitoring precipitation and every 5 to 10 minutes in severe storm warning mode. Early warnings provided by these meteorological radars have also saved many lives.

Finally, daytime and nighttime thermal infrared remote sensor data with high spatial resolution (5-10 m) represent one of the primary methods for obtaining quantitative spatial information on the urban heat island effect (Lo et al., 1997).

Critical Environmental Area Assessment

Urban/suburban environments often include sensitive areas such as wetlands, endangered species habitats, parks, land surrounding treatment plants, and land in urbanized watersheds that provides the runoff for potable drinking water. Relatively stable sensitive environments need be monitored only every 1 to 2 years using a multispectral remote sensor collecting data with a resolution of 1-10 m. For extremely critical areas that could change rapidly, multispectral remote sensors (including a thermal infrared band) should obtain data with a resolution of 0.3-2 m every 1 to 6 months.

Disaster Emergency Response

Recent floods (Mississippi River in 1993; Albany, Georgia, in 1994), hurricanes (Hugo in 1989, Andrew in 1991, Fran in 1996), tornadoes (every year), fires, tanker spills, and earthquakes (Northridge in 1994) have demonstrated that a rectified predisaster remote sensing image database is indispensable. The predisaster data need be updated only every 1 to 5 years; however, multispectral data with high spatial resolution (1-5 m) should be obtained if possible.

When disaster strikes, high-resolution (0.5-2 m) panchromatic and/or near-infrared data should be acquired within 12 hours to 2 days. If the terrain is shrouded in clouds, imaging radar may provide the most useful information. Postdisaster images are registered to the predisaster images, and manual or digital change detection is performed (Jensen, 1996). If precise, quantitative information about damaged housing stock, disrupted transportation arteries, the flow of spilled materials, and damage to above-ground utilities is required, it is advisable to acquire postdisaster panchromatic and near-infrared data with a resolution of 0.3-1 m within 1 to 2 days. Such information was indispensable in assessing damages and allocating scarce cleanup resources during Hurricanes Hugo, Andrew, and Fran (Wagman, 1997) and the recent Northridge earthquake.

USE OF REMOTE SENSING FOR FORECASTING URBAN RESIDENTIAL EXPANSION

The study of residential expansion has a long history that is closely linked to early models of the internal structure of cities in which an urban area is viewed as a series of concentric rings, sectors, or multiple nuclei (Harris and Ullman, 1945).

In those early models, the rate of expansion of the city was treated as a struggle between a series of centrifugal and centripetal forces. In recent decades, researchers have attempted to model this process using empirical data.

Models ranging from those that emerged from urban ecology literature in the mid-1920s to those based on urban economics of the 1960s, which were founded in rent theory. The general assumption of these models was that land prices would be highest in the center of the city where accessibility was greatest. Wealthier and more mobile residents would trade off accessibility for more space at the periphery, while poorer residents would live near the center of the urban area at higher densities. This model was articulated by Alonso (1964), but was directly related to much earlier work on agricultural land-use theory.

An important model of residential growth was developed by Chapin and Weiss (1968). This model was based on the concept of priming actions that trigger secondary actions and together produce land development. This residential location model was designed to allocate residential units to areas experiencing growth.

Recent research by Batty and Longley (1994) has taken a fresh look at these urban models and attempted to rework them in light of emerging research in fractal geometry. Most models of urban growth have modeled this diffusion process as a manifestation of random events; however, it is likely that dynamic urban systems are not random, but deterministic in nature. Therefore, a good time series of events that can capture the underlying patterns needs to be established. Remote sensing that can monitor changes approximately every 3 weeks can provide data not available from traditional sources, such as the decennial census of population and housing.

In light of previous attempts to model residential expansion, a major research effort funded by the National Aeronautics and Space Administration was undertaken. This effort focused on the development of an integrated remote sensing and GIS model that could be used to predict urban expansion between census periods (Jensen et al., 1994). This model was based on a systematic method of capturing and analyzing a wide range of data sources that are indicators of urban development. Unlike previous residential models, this model incorporated census data, land use/land cover, raster-based satellite imagery, building permit data, and postal code geography.

An important goal of the model was to forecast not only future growth patterns, but also the specific number of new single-family homes that might be constructed. To accomplish this goal, it was necessary to measure available land, density of housing units, and residential growth rates. The first component of the model was estimation of the change in the amount of available land for development. This was done using USGS land-use/land-cover polygons from 1976 and SPOTJ classified multispectral imagery from 1989. A land-use change detection resulted in a data set showing land that was urban in 1976 and land that had been converted to urban by 1989 (Figure 8-4). The land-use data provide a basis for

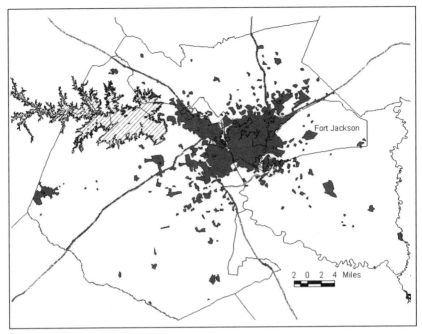

a

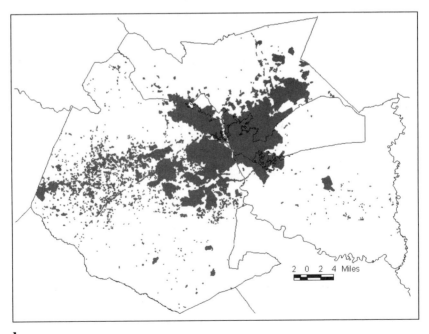

b

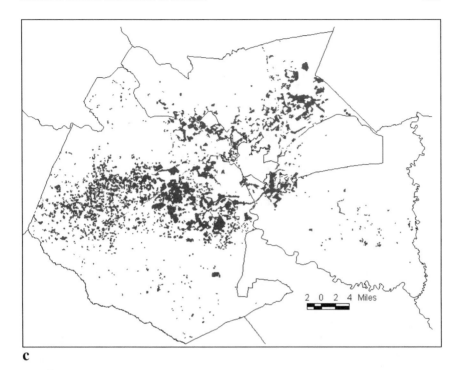

c

FIGURE 8-4   Land-use maps for Columbia, South Carolina, based on (a) 1976 USGS
land-use and land-cover data, and (b) 1990 SPOT 20 × 20 m multispectral data. Shaded
areas indicate urban land uses.   (c) Land converted to urban uses 1976-1990 (shaded
areas).

determining where development can be expected to occur.   In fact, the SPOT
multispectral data at a resolution of 20 × 20 m provided a basis for classifying all
the land in South Carolina as either developable or undevelopable (Plate 8-2,
after page 182).   Developable land consisted of agricultural land, scrub/shrub,
and forests.   Undevelopable land consisted of bodies of water, wetlands, and
publicly owned lands such as parks and military installations.   This approach
provided a useful static view of potential areas of development throughout the
state.   Furthermore, government-owned lands and other protected areas that had
been identified by the Bureau of the Census were identified in the Topographi-
cally Integrated Geographic Encoding Reference (TIGER) line files, and those
land-use polygons were extracted from the developable land areas to present a
more realistic estimate of the amount of land available for development.

Statewide analysis of the 20 × 20 m multispectral data is not economically
feasible on a regular basis.   However, our research indicates that the higher-
resolution 10 × 10 m panchromatic band of SPOT data can be used to detect

changes on a local level. For example, land-cover changes in the Columbia, South Carolina, metropolitan area have been monitored on a 2-year cycle for the past 8 years. These remotely sensed data were integrated with data on 1990 developable land to provide timely updates that clearly identify where disturbances are occurring (see Figures 8-5 and 8-6). This type of synoptic view is more efficient than traditional windshield surveys and much less expensive than aerial photography missions.

Once a measure of the amount of developable land had been determined, it was necessary to estimate the average amount of land per housing unit. This component of the residential forecasting model was calculated for each block group on the basis of the 1990 Census of Housing figures at the block level. The average lot size was adjusted on the basis of the actual urban land use, not the

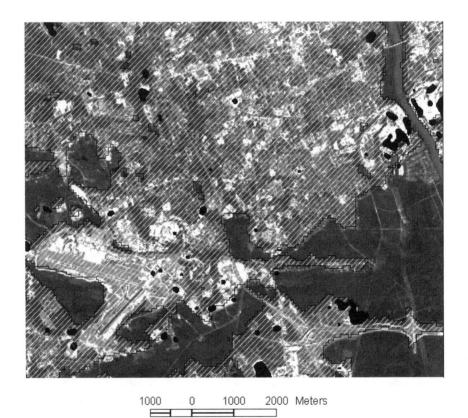

1000   0   1000   2000 Meters

FIGURE 8-5   SPOT 10 × 10 m data overlaid with developed land (diagonal lines). Note the airport and the Interstate highway system.

Landsat MSS Data: April 10, 1973

Landsat MSS Data: March 26, 1981

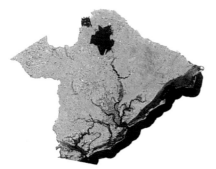
Landsat TM Data: November 9, 1982

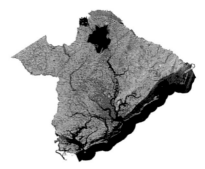
Landsat TM Data: February 3, 1994

PLATE 8-1   Level I urban built-up land (red) extracted from Landsat MSS data (80 x 80 m) from 1973 and 1981 and Landsat TM data (30 x 30 m) from 1982 and 1994 for the Berkeley/Charleston/Dorchester counties of coastal South Carolina.

LEGEND

- ■ Urban/Built-Up Land
- □ Agriculture
- ■ Scrub/Shrub
- ■ Forest
- □ Water
- ■ Saturated Bottomland
- ■ Nonforested Wetland
- ■ Barren/Disturbed Land

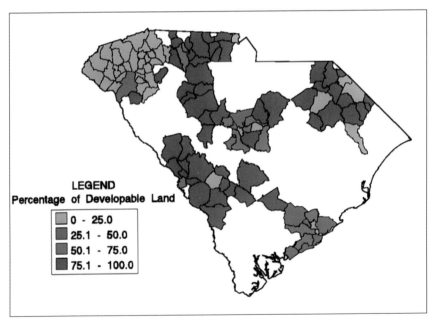

LEGEND
Percentage of Developable Land

- □ 0 - 25.0
- ■ 25.1 - 50.0
- ■ 50.1 - 75.0
- ■ 75.1 - 100.0

PLATE 8-2   Regional land absorption model for South Carolina. Top: Eight land-use classes from SPOT data. Bottom: Percentage of land classified as developable within selected telephone wire centers.

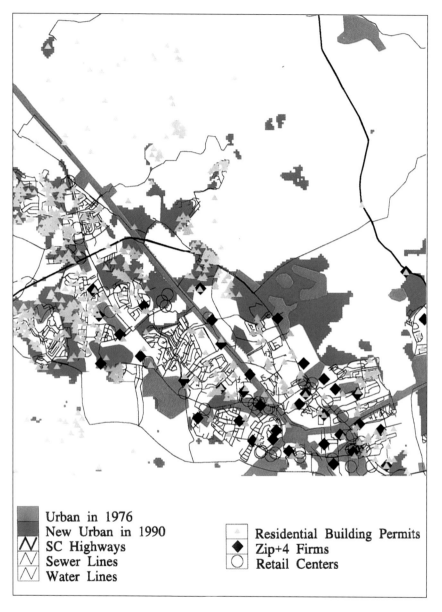

Urban in 1976
New Urban in 1990
SC Highways
Sewer Lines
Water Lines

Residential Building Permits
Zip+4 Firms
Retail Centers

PLATE 8-3   Detailed look at the various data components of the model.

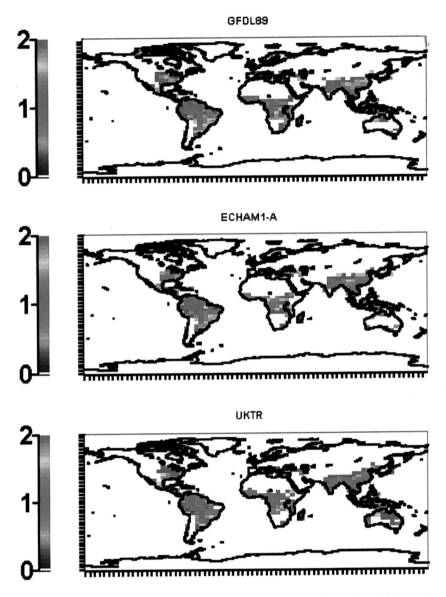

PLATE 10-1   Ratio of average annual potential malaria risk (*Plasmodium falciparum*) under three climate change scenarios to the risk with baseline climate. Based on climate patterns generated by the GFDL89, ECHAM 1-A, and UKTR general circulation models calculated from monthly temperature and precipitation. Global mean temperature increase according to the three scenarios is 1.16°C. Figure prepared by Dr. Pim Martens, Dutch National Institute of Health and Environment Protection. SOURCE: Martens et al. (1997). Reprinted with kind permission from Kluwer Academic Publishers and Pim Martens.

FIGURE 8-6   SPOT 10 × 10 m derived urban land-use changes within census block group polygons derived from SPOT 10 × 10 m data. 1990-1992 changes are in dark gray; 1992-1994 changes are in light gray.

total land area of the block group. Therefore, it could be assumed that the figure was representative of the average lot size in that neighborhood at that time.

The final factor in the model required a spatially detailed empirical estimate of the residential growth rate throughout the metropolitan area. The best indicator of housing changes was a record of 15,303 new single-family building permits for an 11-year period between 1980 and 1990. The permits were geocoded and assigned to block groups. The result was a time-series database for estimating the rate of land-use conversion for small areas. The building permits represent an excellent resource for analyzing spatiotemporal change. The pattern can also be treated as a demographic process—a series of births, deaths, and migrations that result in a changing spatial point pattern. From the viewpoint of real estate developers, the housing market progresses through a life cycle that involves the density of houses, or lot size, and the availability of land, or land absorption rate. When all of these data sources are combined, it is possible to visualize the series of events occurring within an urban area with a considerable amount of detail (Plate 8-3, after page 182).

Rather than trying to fit a generalized expansion model to the entire region,

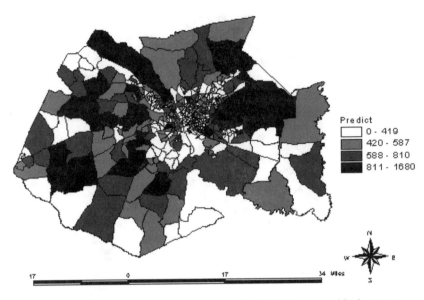

FIGURE 8-7   Number of new houses forecasted in 2005 for census block groups.

it was deemed more useful to summarize the temporal aspects of the permit data for each of the 393 block groups. A regression analysis was performed for each block group to determine the relationship between the number of building permits and time. The parameters of the regression models (slope and intercept) actually became attributes of the polygons. The Y intercept represented an estimate at the start of the study period. The regression coefficient provided a summary of the rate and direction of change throughout the period. The correlation coefficient measured the strength of the trend. With the parameters for the regression model for each polygon, this model has the ability to predict the number of housing units through time. By incorporating the availability of land and housing unit size into the model, it is possible to estimate the period when developable land within a block group will become saturated. These models were used to forecast the number of housing units, the amount of available land, and the year of saturation for the years 1992 through 2005 (see Figure 8-7). This effort lays the foundation for future models that will incorporate spatial information extracted from remotely sensed data (Halls et al., 1994).

CONCLUSIONS

The future interface between social science research and remote sensing depends on what kind of features can be detected and how often the data can be obtained. Remote sensing may be used not only for monitoring change, but also for conducting surveillance. For example, it may become possible not just to count houses, but to count the number of stories and detect changes in structures. Thus remotely sensed data may provide the ability to check on building regulations. It may also be possible to develop some new surrogates for socioeconomic conditions. For example, factors such as lot size, the condition of lawns, numbers of swimming pools, and numbers of vehicles may be used to provide insight into residential quality. These capabilities will also provide extremely valuable inputs for models of residential water usage and the demand for other public services.

Table 8-1 and Figure 8-1 reveal that there are a number of remote sensing systems that currently provide some of the desired urban/socioeconomic information when the required spatial resolution is poorer than 5×5 m and the temporal resolution is 1 to 55 days. However, data with very high spatial resolution ($< 1 \times 1$ m) are needed to satisfy many of the requirements for socioeconomic data. In fact, as shown in Figure 8-1, the only sensor that currently provides such data on demand is aerial photography (0.3-0.5 m). Neither EOSAT/ Space Imaging (1997) with its 1×1 m panchromatic data nor EarthWatch Earlybird (1997) with its proposed 3×3 m panchromatic data nor Quickbird with its 0.8-0.8 m data will satisfy all of the data requirements. None of the sensors, except repetitive aerial photography, can provide the 5-60 minute temporal resolution needed for traffic and parking studies. It may be necessary to have satellite remotely sensed data with higher spatial resolution (0.3-0.5 m) and temporal resolution (1-3 days) to provide much of the desired detailed urban/suburban socioeconomic information, or to utilize aerial photography. Fortunately, the GOES constellation (East and West) and the European Meteosat provide sufficient national and regional weather information at reasonable temporal (30 minutes) and spatial (1-8 km) resolution. Ground-based Doppler radar provides sufficient spatial (4×4 km) and temporal (5-30 minutes) resolution for precipitation and intense storm tracking.

Finally, while remote sensing provides a valuable way to monitor changes on the earth's surface, it can only suggest details about human activity. To obtain this type of information, it is necessary to have a source of data for monitoring the movement of people; consumer behavior; and a wide range of events relating to crime, health, and other matters. It is also important to note that while such knowledge helps route emergency vehicles to our houses and can help utility companies and urban planners prepare for future developments, there is no question that our individual rights to privacy may be jeopardized. It is clear that improvements in the spatial and spectral resolution of sensing systems have the potential to impact our privacy by providing public and private organizations with visual clues regarding the activities in our houses or on our property. From a social science perspective, we will soon have the ability to monitor human activity much more closely than has been possible to date. The key question is whether this type of information can be used to create more efficient urban environments and provide for a more equitable distribution of resources and services.

REFERENCES

Alonso, W.
  1964   *Location and Land Use.* Cambridge, Mass.: Harvard University Press.
Anderson, J.R., E.E. Hardy, J.T. Roach, and R.E. Witmer
  1976   A Land Use and Land Cover Classification System for Use with Remote Sensor Data
          U.S. Geological Survey Professional Paper 964. USGS, Washington D.C.
Angelici, G. L., N.A. Bryant, R.K. Fretz, and S.Z. Friedman
  1980   *Urban Solar Photovoltaics Potential: An Inventory and Modeling Study Applied to the
          San Fernando Valley Region of Los Angeles.* JPL Report. 80-43, Pasadena, Calif.: Jet
          Propulsion Laboratory.
Batty, M., and P. Longley
  1994   *Fractal Cities: A Geometry of Form and Function.* London, England: Academic Press.
Branch, M.C.
  1971   *City Planning and Aerial Information.* Cambridge, Mass.: Harvard University Press.
Chapin F.C., and S.F. Weiss
  1968   A probabilistic model for residential growth. *Transportation Research* 2:375-390.
Clayton, C., and J.E. Estes
  1980   Distributed parameter modeling of urban residential energy demand. *Photogrammetric
          Engineering and Remote Sensing* 45:106-115.
Colcord, J.E.
  1981   Thermal imagery energy surveys. *Photogrammetric Engineering and Remote Sensing*
          47:237-240.
Cowen, D., J. Jensen, J. Halls, M. King, and S. Narumalani
  1993   Estimating Housing Density with CAMS Remotely Sensed Data, *Proceedings, ACSM/
          ASPRS* :35-43.
Cowen, D., J.R. Jensen, G. Bresnahan, D. Ehler, D. Traves, X. Huang, C. Weisner, and H.E. Mackey
  1995   The design and implementation of an integrated GIS for environmental applications. *Pho-
          togrammetric Engineering and Remote Sensing* 61(11):1393-1404.
Eliasson, I.
  1992   Infrared thermography and urban temperature patterns. *International Journal of Remote
          Sensing* 13(5):869-879.
Federal Geographic Data Committee
  1997   *Draft Geospatial Positioning Accuracy Standards.* Washington, D.C.: Federal Geographic
          Data Committee.
Ford, K.
  1979   *Remote Sensing for Planners.* Rutgers: State University. of New Jersey.
Green, K., D. Kempka, and L. Lackey
  1994   Using remote sensing to detect and monitor land-cover and land-use change. *Photogram-
          metric Engineering and Remote Sensing* 60(3):331-337.
Green, N.E.
  1957   Aerial photographic interpretation and social structure of the city. *Photogrammetric Engi-
          neering* 23:89-96.
Gugan, D.J., and I.J. Dowman
  1988   Topographic mapping from SPOT imagery. *Photogrammetric Engineering and Remote
          Sensing* 54(10):1409-1404.
Haack, B., S. Guptill, R. Holz, S. Jampoler, J. Jensen, and R. Welch
  1997   Urban analysis and planning. Pp. 517-553 in *Manual of Photographic Interpretation.*
          Bethesda, Md.: American Society for Photogrammetry and Remote Sensing.
Halls, J.N., D.J. Cowen, and J.R. Jensen
  1994   Predictive spatio-temporal modeling in GIS. Pp. 431-448 in *Advances in GIS Research,*
          Vol. 1, T.C. Waugh and R.G. Healy, eds., London, England: Taylor and Francis.

Harris, C.D., and E.L. Ullman
  1945   The nature of cities. *Annals of the American Academy of Political and Social Science* 244:7-17.
Henderson, F.M., and J.J. Utano
  1975   Assessing general urban socioeconomic conditions with conventional air photography. *Photogrammetria* 31:81-89.
Holz, R.K.
  1988   Population Estimation of Colonias in the Lower Rio Grande Valley Using Remote Sensing Techniques. Paper presented at the annual meeting of the Association of American Geographers, Phoenix, AZ.
Jadkowski, M.A., P. Convery, R.J. Birk, and S. Kuo
  1994   Aerial image databases for pipeline rights-of-way management. *Photogrammetric Engineering and Remote Sensing* 60(3):347-353.
Jensen, J.R., ed.
  1983a Urban/suburban land use analysis. Pp. 1571-1666 in *Manual of Remote Sensing*, 2nd ed., R.N. Colwell, ed. Falls Church, Va.: American Society of Photogrammetry.
Jensen, J.R.
  1983b Detecting residential land-use development at the urban fringe. *Photogrammetric Engineering and Remote Sensing* 48:629-643.
  1995   Issues involving the creation of digital elevation models and terrain corrected orthoimagery using soft-copy photogrammetry. *Geocarto International: A Multidisciplinary Journal of Remote Sensing* 10(1):1-17.
  1996   *Introductory Digital Image Processing: A Remote Sensing Perspective.* Saddle River, N.J.: Prentice-Hall.
Jensen, J.R., D.C. Cowen, J. Halls, S. Narumalani, N. Schmidt, B.A. Davis, and B. Burgess
  1994   Improved urban infrastructure mapping and forecasting for BellSouth using remote sensing and GIS technology. *Photogrammetric Engineering and Remote Sensing* 60(3):339-346.
Jensen, J.R., X. Huang, D. Graves, and R. Hanning
  1996   Cellular phone transceiver site selection. Pp. 117-125 in *Raster Imagery in Geographic Information Systems*, S. Morain and S. Baros, eds. Santa Fe, NM: OnWard Press.
Lacy, R.
  1992   South Carolina finds economical way to update digital road data. *GIS World* 5(10):58-60.
Leberl, F.W.
  1990   *Radargrammetric Image Processing.* Norwood, Mass.: Artech House.
Lindgren, D.T.
  1985   *Land Use Planning and Remote Sensing.* Boston, Mass.: Martinus Nijhhoff Inc.
Lo, C.P.
  1986   The human population. Pp. 40-70 in *Applied Remote Sensing*, New York: Longman.
  1995   Automated population and dwelling unit estimation from high-resolution satellite images: A GIS approach. *International Journal of Remote Sensing* 16(1):17-34.
Lo, C.P., D.A. Quattrochi, and J.C. Luvall
  1997   Application of high-resolution thermal infrared remote sensing and GIS to assess the urban heat island effect. *International Journal of Remote Sensing* 18(2):287-304.
McCoy, R.M., and E.D. Metivier
  1973   House density vs. socioeconomic conditions. *Photogrammetric Engineering* 39:43-49.
Mintzer, O.W., ed.
  1983   Engineering applications. Pp. 1955-2109 in *Manual of Remote Sensing*, 2nd ed., R.N. Colwell, ed. Falls Church, Va.: American Society for Photogrammetry.

Monier, R.B., and N.E. Green
   1953   Preliminary findings on the development of criteria for the identification of urban struc-
           tures from aerial photographs. *Annals of the Association of American Geographers,*
           special issue.

Montesano, A.P.
   1997   Roadmap to the future. *Earth Observation* 6(2):16-19.

Morain, S.A., and A.M. Budge, eds.
   1996   Earth observing platforms and sensors. Vol. 2 in *Manual of Remote Sensing, 3rd Edition,*
           A. Ryerson, editor-in-chief. CD-ROM. Bethesda, Md.: American Society for Photo-
           grammetry and Remote Sensing.

Petrie, G., and T.J.M. Kennie
   1990   *Terrain Modeling in Surveying and Civil Engineering.* London, England: Whittles Pub-
           lishing.

Philipson, W.
   1997   *Manual of Photographic Interpretation.* Bethesda, Md.: American Society for Photo-
           grammetry and Remote Sensing.

Raymondo, James C.
   1992   *Population Estimation and Projection: Methods for Marketing, Demographic, and Plan-
           ning Personnel.* New York: Quorum Books.

Schultz, G.A.
   1988   Remote sensing in hydrology. *Journal of Hydrology* 100:239-265.

Stoeckeler, E.G.
   1979   Use of aerial color photography for pavement evaluation studies. *Highway Research
           Record* 319:40-57.

Tobler, W.
   1969   Satellite confirmation of settlement size coefficients. *Area* 1:30-34.

Tuyahov, A.J., C.S. Davies, and R.K. Holz
   1973   Detection of urban blight using remote sensing techniques. *Remote Sensing of Earth
           Resources* 2:213-226.

Wagman, D.
   1997   Fires, hurricanes prove no match for GIS. *Earth Observation* 6(2):27-29.

Warner, W.S., R.W. Graham, and R.E. Read
   1996   Urban survey. Pp. 253-256 in *Small Format Aerial Photography.* Scotland: Wittles Pub-
           lishing.

Wolff, R.S., and L. Yaeger
   1993   *Visualization of Natural Phenomena.* Santa Clara, Calif.: Telos Springer-Verlag.

9

Social Science and Remote Sensing in Famine Early Warning

Charles F. Hutchinson

The specter of famine has haunted humans throughout history. Yet despite the havoc it has wreaked on various populations, the causes of famine are poorly understood. Famine has been viewed as the will of God, as an accident of nature, as an inexorable consequence of economics and, most recently, as a social process. Because it strikes a fundamental chord that links all of humanity, famine, or the threat of famine, is an important issue that occupies a more prominent place in popular thinking than issues such as high rates of malnutrition, morbidity, and mortality in the low-income countries of the world. Thus, it is generally accepted that there is a need to better understand the process of famine so it may be averted and ultimately eliminated.

Formal efforts to monitor food security and predict famine began more than a century ago in India, where colonial administrators sought evidence of impending food shortages in rural areas to avert disaster. While the resulting Famine Codes have persisted there, it was not until the great droughts of the late 1960s and early 1970s in the Horn and Sahel regions of Africa that the international community paid attention to the need to develop a more comprehensive system for routine monitoring of food security (Arnold, 1988).

Between 1975 and the present, several systems have evolved for monitoring food conditions, primarily in Africa. An increasingly sophisticated array of tools has been applied to the problem, including a number of satellite remote sensing techniques (Hutchinson, 1991). However, the utility of these new tools has been hampered by inadequacies in (1) general understanding of the famine process, and (2) specific understanding of the interaction between restrictions in access to food and reactions to those restrictions among various groups.

The objectives of this chapter are to review different models of famine and how they have been adopted in various famine early warning systems, to consider the evolving role of remote sensing in famine early warning, and to explore how social science and information technologies (i.e., remote sensing and geographic information systems) appear to be converging in the famine early warning arena as a potential model for other integrated assessments.

PRACTICAL IMPLEMENTATION OF FAMINE MODELS

The major systems for monitoring food security—or the likelihood of famine—operate at continental and global scales. The monitoring activities described here rely on secondary aggregate data (data routinely gathered and reported for administrative districts by national governments) and on observations made by satellites. Because of the expense involved, primary field data are gathered only after a country or region within a country has been identified as a potential problem.

The Food and Agriculture Organization (FAO) of the United Nations launched the Global Information and Early Warning System (GIEWS) in 1975. At that time, famine was defined in Malthusian terms as a situation in which demand for food exceeds supply (i.e., failure of supply). The tool developed to describe food status was a national "food balance sheet" in which food demand as a function of population size was weighed against agricultural production for the current year, plus scheduled imports and food stocks carried over from previous years. Although fairly crude, the food balance sheet came into general use because it could be developed relatively early in the growing season and yielded a product that was easily understood and acted upon (Global Information and Early Warning System, 1997).

The return of catastrophic droughts in the Horn and Sahel regions of Africa in 1984 again drew international attention to the problem of famine. At the Bonn Summit of 1985, donor countries agreed to renew their collective efforts to detect and avert future famine, particularly in Africa. This renewed interest spurred developments in famine early warning, and efforts were made to improve the food balance sheet approach pioneered by GIEWS. Almost all suggested refinements to the approach involved improving the supply side of the balance sheet, and national estimates of crop production became the obvious target. Two general approaches to improving crop production estimates emerged. One was based on models of crop yield driven by direct observations of precipitation. Although this approach was promising, its adoption was hindered by a lack of reliable meteorological data and an insensitivity to other factors that might affect total production (e.g., availability of inputs, estimates of planted area). The other, more direct approach was based on observing the progress of the growing season using remotely sensed data (Hutchinson, 1991).

Beginning in 1979, great strides were made in regional monitoring based on the use of satellite data, especially those provided by the National Oceanic and

Atmospheric Administration (NOAA) series of satellites carrying the Advanced Very High Resolution Radiometer (AVHRR) instrument. With these data it was possible to monitor conditions routinely for all of Africa, summarized over a standardized 10-day period (dekad), for pixels with an area of about 16 km^2. While this coarse scale does not allow statements to be made about specific fields or communities, it does provide an overview of how the growing season is progressing for an area that may contain several communities or for an administrative district. Similarly, it is difficult at such a scale to derive quantitative estimates of crop yields for specific areas. However, the development of an archive of observations since 1981 has made it possible to compare a given period of the current year with the same period during the past year or with mean conditions over the period of record, and thus to develop a reasonably reliable estimate of the quality of the growing season and anticipated crop yield. Thus, it is possible to assert with some confidence whether the cropping season in an area will be better or worse compared with last year or the average.

Two programs to incorporate satellite data in famine early warning were launched in 1985: FAO developed the Africa Real Time Environmental Monitoring Information System (ARTEMIS) to provide data to GIEWS and other regional monitoring programs (e.g., locust control operations) for all of Africa (Hielkema et al., 1986), while the U.S. Agency for International Development (USAID) developed the Famine Early Warning System (FEWS), which operated initially in the Sahel and Horn of Africa (Hutchinson, 1991). Both programs use and share AVHRR data processed by the National Aeronautics and Space Administration (NASA) to describe vegetation conditions, and both have added rainfall estimates based on cloud observations derived from the European Meteosat satellite (Snijders, 1991).

Despite improved estimates of food production from satellite data, it became obvious that famine was more complex than a simple failure of food supply. Studies of famine during the 1970s revealed that food was available during these events, yet people still starved (Garcia, 1981). It was proposed that in many, if not most, emergency situations, food may be available, but the mechanisms of exchange (entitlements) by which people have traditionally gained access to food cease to function (Sen, 1981). Thus, rather than a failure of supply, famine is caused by a failure of effective demand.

The need to gather, monitor, and analyze data on food access presented new challenges. FEWS initially adopted a "convergence-of-evidence" approach used routinely in air photo and satellite image interpretation (Estes et al., 1983). In addition to information derived from images, this approach considers a wide array of data types, with an interpretation being made that is supported by all the data. In the FEWS approach, a minimum of three sets of information, or classes of "indicators," was considered. The first set included those factors that might suggest food supply (e.g., observed precipitation; early season yield estimates; reports of the incidence of pests, such as locusts; and estimates of food stocks in granaries carried over from the preceding year). The second set comprised those factors that might suggest food access (e.g., cereal prices; small animal prices; terms of trade between cereal and animals; labor prices; and anecdotal information, such as reports by travelers about conditions in outlying areas). The third set included those factors that might indicate levels of development (e.g., access to roads, water, and health and school facilities). These sets of data were then interpreted together. For example, if the growing season was below average for a district, but stocks were high, the area was accessible by road, and terms of trade held steady, convergence of evidence suggested that an emergency was unlikely. Conversely, if an indifferent year followed 2 years of poor harvest in an isolated area and terms of trade began to deteriorate quickly, convergence of evidence suggested that the poorest households would have restricted access to food, and the likelihood of an emergency was high.

Adoption of the convergence-of-evidence approach was an advance in early warning because it increased confidence in the assessment. It also offered a number of other benefits because it led to the development of large, dynamic, general-purpose databases that have application beyond early warning. For example, the numbers of health care clinics or schools in a district suggested levels of development that might also be used to infer ease of access to food. Similarly, distance to primary roads would indicate the level of development and thus access to markets or alternative sources of income. While indicative of vulnerability to food security emergencies, these same data could be used as guidelines for long-term development (e.g., to determine where clinics, schools, or roads were most needed).

While the convergence-of-evidence approach offered a relatively sensitive and reliable indication of current conditions, it still provided little information on the differential impacts an emergency might have on various groups or types of households (e.g., pastoralists, subsistence farmers, female-headed households). Gaining an understanding of the economic and social contexts in which an emergency, such as a drought, might play out at the household level became and remains a significant challenge.

Efforts continued to focus on gathering and analyzing secondary aggregate data, but there was a gradual shift from the simple convergence-of-evidence approach to an approach that attempted to interpret data with regard to how they might reflect household response to current conditions. Watts (1983) and later Corbett (1988) offered a simple conceptual model in which the way households react to emergencies or perceived threats to their economic condition is determined by considering how they commit household resources. A modification of this model was adopted by FEWS; see Figure 9-1 (Famine Early Warning System, 1997). In this model, famine is viewed as a process that unfolds over time. Monitoring data can be placed within this framework so that conditions at the household level can be inferred from aggregate data. For example, terms of trade can be expected to deteriorate fairly early in an emergency and continue to

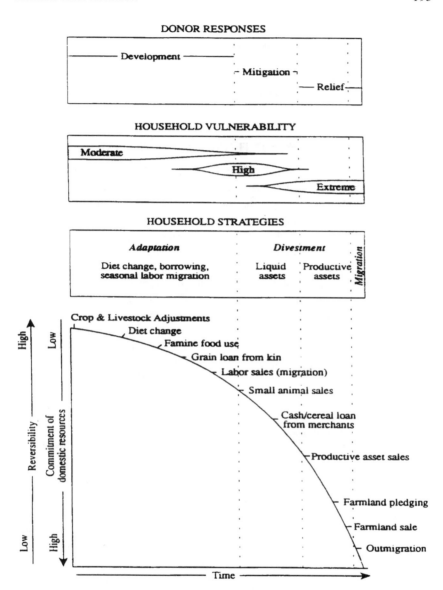

FIGURE 9-1   Household responses to the threat of food security emergencies.

decline as grain prices rise because of demand and as the relative prices of small animals (e.g., goats) fall because more are offered for sale. Further, it can be inferred that an emergency is quite serious when resources that are critical to household livelihoods (e.g., draft animals, farm implements) begin to appear in markets, and that a crisis has arrived when large-scale migrations to urban areas and/or refugee centers begin to take place.

OPPORTUNITIES FOR INTEGRATION OF SOCIAL SCIENCE

Continued work on the basic nature of vulnerability to famine (Watts and Bohle, 1993) and how this vulnerability is manifested among various groups (Seaman, 1997) has opened opportunities for further integration of remote sensing and social science in famine early warning. Certainly there have been significant advances in the power with which aggregate data (e.g., district-level production data) can be manipulated—for example, using geographic information systems—and in the ways these data can be combined with other spatially referenced observations (e.g., subdistrict-level land use derived from remote sensing data) to yield more geographically specific results. Nevertheless, there are very real limits on inference that cannot be overcome without a better understanding of the economic, social, and political differences among groups and households, and how these differences might vary in space and time. Few of these conditions can be revealed through remote sensing. In recognition of these limitations, FEWS has added some caveats to the interpretation of Figure 9-1 (Famine Early Warning System, 1997:12):

> The ability of households to withstand emergencies is conditioned by (a) the depth, diversity and quality of their resource base, (b) the breadth of their income portfolio, and (c) their relationship to economic, social and political hierarchies.

Consequently, in parallel with the efforts of FEWS to refine the use of quantitative aggregate data for inferring household responses to and the impacts of emergencies (a top-down approach), others in the early warning community have sought to develop a more basic qualitative understanding of household economies among various livelihood systems and wealth groups that might be extended over an area (a bottom-up approach). The Save the Children Fund (SCF) has developed a "food economy" approach based on structured interviews with key informants within "food economy zones" (Seaman, 1997). The resulting data are entered into a computerized model, which can then be subjected to various perturbations (e.g., simulation of food production losses through drought). In this way, an estimate of the impact on "household food income" is projected. This estimate is expressed differentially by wealth type of household, and the net food shortfall is mapped out in relation to the food economy zone. Thus where

FEWS, by necessity, applies a generic model of household behavior in all areas and to all groups, SCF develops a set of models that is area- and group-specific.

The bottom-up approach based on specific household models also has its limitations, however. Models developed through field survey are essentially static. If significant changes occur within a region (e.g., if a new road is built or a mine is opened), the models must be revised, again through resource-intensive field survey. Also, the system is usually geographically restricted in its application: it tends to be used in designated areas where relief operations are planned, with less attention being paid to areas outside these zones, making it difficult to extend the findings over larger (e.g., multiethinic, multinational) areas. Finally, the numbers yielded by this approach are often viewed as "soft" in that they incorporate a good deal of qualitative or nonformal data derived from interviews with key informants, rather than a statistically rigorous sample.

As a result, the top-down and bottom-up approaches have been judged to be complementary rather than competitive (Global Information and Early Warning System, 1997). The maintenance of dynamic national databases makes it possible to make statements about an entire country, and offers benefits that extend beyond famine early warning. Moreover, the limitations of a single, universal model of household response resulting from the restricted use of aggregate data can be overcome by developing region- and group-specific models through field survey. With a more specific understanding of household behaviors, it will be possible to offer differential interpretations of the effects of an emergency on different households and different areas using the same monitoring data.

At present, discussion of the involvement of social science in the food security community is restricted primarily to what the fields of anthropology and geography might contribute to an improved understanding of household behavior, largely through the use of rapid qualitative field survey methods combined with the use of geographic information systems to provide a link to routinely reported aggregate data. However, the potential contribution of social science extends beyond this level to issues of theory and the use of more quantitative techniques to achieve better understanding and prediction of human behavior. The need for collaboration with the social sciences has been acknowledged and will occur, at least within this initial context (Global Information and Early Warning System, 1997). Once the fundamentally social nature of the famine process is acknowledged, expansion of this collaboration will be recognized as the most fruitful frontier for early warning research. The greatest challenges to such increased collaboration may well lie in bridging the cultural gaps that exist between academic disciplines. The famine early warning community has recognized these challenges and the need to resolve them.

REFERENCES

Arnold, D.
   1988   *Famine: Social Crisis and Historical Change.* Oxford, England: Basil Blackwell.
Corbett, J.
   1988   Famine and household coping strategies. *World Development* 16(9):1099-1112.
Estes, J.E., E.J. Hajic, and L.R. Tinney
   1983   Fundamentals of image analysis: Analysis of visible and thermal infrared data. Pp. 987-1124 in *Manual of Remote Sensing, Second Edition,* R.N. Colwell, ed. Washington, D.C.: American Society of Photogrammetry.
Famine Early Warning System (FEWS)
   1997   Vulnerability analysis and FEWS. Annex 3/2 *in Global Information and Early Warning System (GIEWS). Summary Report: Second Informal Meeting on Methodology for Vulnerability Assessment.* ES:GCP/INT535/EEC. Rome, Italy: Food and Agriculture Organization of the United Nations.
Garcia, R.V.
   1981   *Drought and Man: The 1972 Case History, Vol. 1. Nature Pleads Not Guilty.* Oxford, England: Pergamon Press.
Global Information and Early Warning System (GIEWS)
   1997   *Summary Report: Second Informal Meeting on Methodology for Vulnerability Assessment.* ES:GCP/INT535/EEC. Rome, Italy: Food and Agriculture Organization of the United Nations.
Hielkema, J.U., J.A. Howard, C.J. Tucker, and H.A. Van Ingen Shenau
   1986   The FAO/NASA/NLR ARTEMIS system. Pp. 147-160 in *Proceedings: Twentieth International Symposium on Remote Sensing of Environment.* Ann Arbor: Environmental Research Institute of Michigan.
Hutchinson, C.F.
   1991   Uses of satellite data for famine early warning in sub-Saharan Africa. *International Journal of Remote Sensing* 12(6):1405-1421.
Seaman, J.
   1997   The food economy approach to vulnerability assessment and the RiskMap computer program. Annex 3/1 in *Global Information and Early Warning System (GIEWS). Summary Report: Second Informal Meeting on Methodology for Vulnerability Assessment.* ES:GCP/INT535/EEC. Rome, Italy: Food and Agriculture Organization of the United Nations.
Sen, A.K.
   1981   *Poverty and Famine: An Essay on Entitlement Deprivation.* Oxford, England: Clarendon Press.
Snijders, F.L.
   1991   Rainfall monitoring based on Meteosat data—a comparison of techniques applied to the western Sahel. *International Journal of Remote Sensing* 12:1331-1347.
Watts, M.J.
   1983   *Silent Violence: Food, Famine and Peasantry in Northern Nigeria.* Berkeley: University of California Press.
Watts, M.J., and H.G. Bohle
   1993   The space of vulnerability: The causal structure of hunger and famine. *Progress in Human Geography* 17(1):43-67.

10

Health Applications of Remote
Sensing and Climate Modeling

Paul R. Epstein

Remote satellite sensing of the oceans, land masses, ice cover, and the atmosphere has been used for understanding biogeochemical cycles and their relationships to biotic activity. An important insight emerging from research on climate and ecosystems is that climatic changes and variations alter the ecology of disease organisms that attack human beings and their food supplies. Remotely sensed data being used to monitor climatic phenomena and improve understanding and forecasting of climate are proving useful for forecasting the spread of disease and offer great potential for the development of health early warning systems. Remote sensing can aid in the monitoring of disease carriers (vectors), breeding sites, and animal reservoirs in both marine and terrestrial ecosystems. Integrated into geographic information systems, it can advance disease surveillance, as well as aid in the development of timely, environmentally sound public health interventions.

This chapter examines five applications of remote sensing for disease surveillance: (1) monitoring coastal algal blooms and toxic phytoplankton to support early warning systems for paralytic shellfish poisoning and cholera; (2) monitoring terrestrial habitats to identify and control mosquito and rodent disease vectors; (3) building models of climate variability that can be used to predict conditions conducive to disease outbreaks; (4) using climate-change models to project potential disease distribution; and (5) detecting tropospheric temperatures to help understand physical and biological changes, particularly the spread of disease, at high altitudes.

MONITORING COASTAL ALGAL BLOOMS

Remote sensing data from the Coastal Zone Color Scanner (CZCS) and the Sea-viewing Wide-Field-of-view Sensor (SeaWiFS), as well as measurements of sea-surface temperatures using the Advanced Very High Resolution Radiometer (AVHRR), have been used to assess phytoplankton blooms (Feldman et al., 1984; Brock et al., 1991; Brock and McClain, 1992) and primary productivity (Aiken et al., 1992). In one application, the AVHRR instrument, which senses red and infrared wavelengths, provides measures of sea-surface temperatures, which are correlated with the appearance of algal blooms. In Woods Hole, Massachusetts, AVHRR images are used in real time to detect and follow spring plumes from Massachusetts rivers flowing to the coast. Based on these images, boats are sent out for targeted sampling to detect blooms of *Alexandrium tamarense*, a toxic phytoplankton responsible for paralytic shellfish poisoning. When *A. tamarense* is found, shellfish beds in the area are closed to harvesting, thus preventing outbreaks of the disease (D. Anderson, Woods Hole Oceanographic Institution, personal communication, 1993).

Similar remote sensing technology can provide early warning of cholera outbreaks. Since the 1960s, researchers in Bangladesh have associated outbreaks of cholera with seasonal coastal algal blooms (Cockburn and Cassanos, 1960). Recently, Colwell and associates have used fluorescent antibody probes to identify a viable, nonculturable "dormant" form of *Vibrio cholerae*, which attaches to a wide range of marine life (Colwell et al., 1985) and reemerges to an infectious state along with algal blooms. Algal blooms are grazed by zooplankton that, over days to weeks, are capable of concentrating *V. cholerae* bacteria to infectious doses in their egg sacs. Zooplankton may be consumed directly in drinking water (as in Bangladesh), or be passed up the marine food chain (through shellfish filter feeders or finfish) and thus be introduced into human populations indirectly.

Consequently, early detection of phytoplankton blooms and targeted sampling for cholera bacteria can constitute a cholera early warning system. Early detection in the marine environment can allow for timely institution of public health measures that include temporary bans on shellfish and finfish consumption, use of new oral recombinant immunizations, increased chlorination of water, and preparation of medical treatment facilities. Early treatment can reduce the case fatality rate from 50 to 1 percent.

We are currently in the Seventh World Pandemic of cholera (El Tor strain). This pandemic began in the Bay of Bengal in the 1960s, arrived in Africa in the 1970s, and infected Latin America in the 1990s. In 1992, a new strain of cholera emerged, called *V. cholerae* O139 Bengal. There is no cross-immunity between the older and newly emerged forms (i.e., previous infection with El Tor does not protect against O139 Bengal). While currently confined to the Bay of Bengal, this new organism could find its way in bilge or ballast water, or with a traveler, to other parts of the world. Thus a cholera early warning system could help anticipate and reduce the impact of an Eighth World Pandemic of cholera.

MONITORING TERRESTRIAL HABITATS
OF DISEASE VECTORS

Remote sensing has been used to delineate the habitats of vectors bearing diseases such as African sleeping sickness (Epstein, Rogers, and Slooff, 1993; Rogers and Packer, 1993) and malaria (Dister et al., 1993) so that controls can be instituted. The potential of the approach can be illustrated with the example of Eastern Equine Encephalitis (EEE), a disease transferred to humans by the bite of *Aedes vexans* mosquitoes. EEE most often affects children, and—even in small outbreaks—the consequences are grave and are terrifying for communities affected. Up to half of those infected may die, and half of the survivors are left with permanent neurological complications.

Knowing where temporary pools of standing water are, when they will appear, and perhaps how long they will last is necessary so that environmentally appropriate actions can be taken to control populations of EEE-infected mosquitoes. Early use of Bti (*Bacillus thuringiensis* var. *israelensis*), a nontoxic, inexpensive larvicide, is the alternative to widespread spraying of the adulticide malathion. Maturation of larvae to adults occurs in about 7 days, so accurate information on standing pools of water within 2 days after a rain will allow time for dip sampling to test for the presence of vectors in pools and subsequent application of Bti (Epstein, Rogers and Slooff, 1993).

The best approach for mapping standing water dependably involves the acquisition of remotely sensed images. These and other data layers can then be used together for analysis in an appropriate geographic information system. Real-time information following summer rains can be obtained from oblique-angled synthetic aperture radar (SAR), which can penetrate vegetative and cloud cover (Imhoff and McCandless, 1988; Imhoff and Gesch, 1990) to distinguish smooth water surfaces, thus helping to focus dip sampling and the application of larvicidal treatment in a timely fashion.

Imagery from Landsat, with 30-m spatial resolution and coverage every 16 days (at best), or from Système pour l'Observation de la Terre (SPOT), which has the advantage of relatively more on-demand coverage and 20- and 10-m spatial resolution, will be helpful in developing a series of baseline maps identifying areas at risk for infection. SAR may be most appropriate for providing real-time accurate estimates of the locations of standing water. Aircraft-collected SAR data could be acquired and processed at an appropriate scale and processed for use in a timely fashion.

MODELING CLIMATE VARIABILITY
AND DISEASE OUTBREAKS

During the warm phase of the El Niño/Southern Oscillation (ENSO), specific areas of the globe are consistently affected by drought, whereas others experience excessive precipitation. While Southeast Brazil has rain, for example, Northeast Brazil has intensified drought. The climatic effects of ENSO are stronger (more consistent) in some areas; Southern Africa repeatedly experiences drought during an El Niño. All tropical oceans warm in relation to the ENSO pattern, and evaporation from the Atlantic can cause floods in a warmer Central Europe.

Tracking ENSO events in relation to epidemics is a key to identifying the impacts of climate variability and weather on disease patterns. Associations in themselves are not proof of causation, but a preponderance of globally distributed evidence and a plausible mechanism (extremes of precipitation and temperature) lend credence to a strong role for climate in disease distribution.

El Niño warm events are associated with upsurges of cholera in Bangladesh (R.B. Sack, The Johns Hopkins University School of Public Health, personal communication); typhoid, shigellosis, and hepatitis after flooding in South America (Cabello, 1991); viral encephalitides (Murray Valley and epidemic polyarthritis from Ross River virus) in Australia (Nicholls, 1993); and Eastern Equine Encephalitis in Massachusetts (Edman et al., 1993). Other ENSO-related outbreaks of disease include malaria worldwide (Bouma et al., 1994a), in Pakistan (Bouma et al., 1994b), and in Venezuela (Bouma and Dye, 1997); malaria and dengue ("breakbone") fever upsurges in Costa Rica (Kassutto and Epstein, unpublished data); dengue fever in Northeast Brazil (unpublished data); epidemic malaria in the Indian subcontinent, 1874-1945 (M.J. Bouma, London School of Hygiene and Tropical Medicine, personal communication); and agricultural rodent infestations in Zimbabwe (1973-1983, 1994) (Epstein and Chikwenhere, 1994). There is also a direct relationship between monsoons, which are biennially related to ENSO, and the spread of the brown plant hopper (*Nilaparvata lugans*) rice pest in Southeast Asia (Walker, 1994).

Interannual climate variability such as ENSO may be related to upsurges of soil-borne organisms as well. From the 428 reported cases of San Joaquim Valley fever (due to the fungus *Coccidioidomycosis immitis*) in the 1980s, 1200 cases occurred in 1991, and over 4000 in the El Niño years of 1992 and 1993 (Centers for Disease Control and Prevention, 1994). An earthquake and a prolonged drought followed by torrential rains are considered to have been contributory factors.

Additionally, disease events across taxa appear to cluster during ENSO warm-phase years. Disease events along the U.S. Atlantic coast during 1987, an ENSO warm-phase year heralding the warmest year of this century to that time, included extensive Caribbean coral bleaching; a large Florida sea grass die-off;

transfer of the agent of neurological shellfish poisoning (*Gymnodinium breve*) from the Gulf of Mexico to North Carolina; a large die-off of sea mammals in New England and the North Sea; the emergence of amnesic shellfish poisoning in Prince Edward Island (caused by a newly discovered diatom toxin, domoic acid, and later appearing worldwide) (Epstein, Ford, and Colwell, 1993); and an outbreak of spruce budworms in the Northeast Canadian balsam forests. ENSO warm events are also correlated with new appearances of harmful algal blooms in Asia (Hallaegraeff, 1993) and along the U.S. Atlantic coast (M. Altalo, Scripps Institution of Oceanography, personal communication).

This evidence of relationships between ENSO and disease outbreaks suggests that predictions of ENSO can be used in health early warning systems. Dynamic atmospheric-oceanic coupled general circulation models that depend in part on remote sensing of sea-surface temperatures provide predictions of ENSO and of its climatic effects. These models and their predictions are based on analysis of geographic patterns of climate since 1877 (Kaladis and Diaz, 1989; Glantz et al., 1991) and are improving in their ability to make regional predictions of climatic events. As the models increase in predictive skill, they will become increasingly useful for predicting climatic conditions conducive to disease outbreaks.

USING CLIMATE CHANGE MODELS
TO PROJECT DISEASE DISTRIBUTION

Animals and plants have clear thresholds for viability, as well as temperature and humidity ranges in which they mature, replicate, and thrive (Gill, 1920a, b; Burgos, 1990; Burgos et al., 1994; Dobson and Carper, 1993). Shifts in temperature isotherms in latitude and altitude with climate change could thus have profound impacts on ecotones (the geographical dividing lines between ecosystems), on biota, and in particular on the distribution of pests and pathogens.

Several models using remote sensing, geographic information systems, and general circulation climate models have been used to project for particular areas of the world how conditions conducive to vector-borne diseases may change with global warming scenarios. The affected diseases include malaria, schistosomiasis (Martens et al., 1994; Matsuoka et al., 1994), and dengue fever (Focks et al., 1995). Plate 10-1 (facing page 183) shows the output of one such model for malaria.

UNDERSTANDING THE SPREAD OF
DISEASE AT HIGH ALTITUDES

Recent reports indicate that malaria and dengue fever are appearing at higher altitudes than at any time during this century. In addition, plants have been observed to be moving to higher altitudes on 30 Alpine peaks, in the U.S. Sierra

Nevada, in Alaska, and in New Zealand. Moreover, summit glaciers are retreating on many continents, and ice caps show evidence of accelerated warming during this century (Thompson et al., 1995).

An initial examination of data from the Microwave Sounding Unit (MSU) and infrared sensors (Susskind et al., 1997) suggests that in El Niño years, warming in the upper atmosphere may exceed warming occurring on the earth's surface. There are several possible contributing factors to explain these observations. One is the increased relative heat absorption of carbon in the upper troposphere, because it is cooler at higher altitudes. Second, increased tropospheric water evaporation due to deep oceanic warming (Southwest Pacific, Atlantic, and Indian Oceans) (Parilla et al., 1994; Thwaites, 1994), exaggerated during El Niño years (Graham, 1995), may augment greenhouse warming and increase high, heat-trapping clouds. Third, sulfur-enriched lower clouds may also increase with increased atmospheric water vapor, and may reflect and absorb solar energy and cool the earth's surface with rain.

Such analyses can help in developing causal models that link predictable seasonal-to-interannual climate changes, such as ENSO and global warming processes, to their effects on the ecology of disease organisms. This sort of understanding will be valuable to public health officials in affected areas, such as major high-altitude tropical cities, in forecasting the potential for epidemics and taking appropriate action.

SUMMARY AND CONCLUSION

The costs of not understanding present climate instability and likely changes in climate due to human activities may be enormous. Disease outbreaks cause disability and mortality, and the impacts can ripple through societies and economies. In 1995, the dengue outbreak cost Central American nations $7.5 million in control efforts alone; Peruvian fisheries lost $750 million in seafood exports during the 1991 cholera epidemic; and international airline and hotel industries lost an estimated $2 billion from the Indian plague in 1994. The global resurgence of malaria, dengue fever, and cholera—and the emergence of relatively new diseases such as Ebola—can impact development, trade and tourism, agriculture, and livestock.

Remote sensing alone or integrated into geographical information systems and general circulation models has multiple applications for understanding biological processes, and in particular, disease phenomena. Health early warning systems that can identify climate conditions conducive to outbreaks and disease clusters are becoming feasible—enabling early, environmentally sound public health interventions. For vector-borne diseases, such interventions include, among others, immunizations, where appropriate; source reduction, such as neighborhood cleanups and the clearing of breeding sites; selective applications of pesticides or the nontoxic Bti; and community distribution of bednets.

A cholera early warning system that uses remote sensing to detect coastal algal blooms and target surveillance has immediate relevance to protecting populations. Public health responses based on these data include increased surveillance, preparation of oral rehydration treatment centers, increased chlorination of water supplies, and temporary closure of shellfish beds and fishing grounds.

Additionally, remote sensing can be used in projecting future potential disease distribution due to climate change. It can also play a central role in a multidisciplinary exploration of current physical and biological changes occurring at high altitudes, thus providing policy makers with important information on climate trends and their impacts.

REFERENCES

Aiken, J., G.F. Moore, and P.M. Holligan
    1992   Remote sensing of oceanic biology in relation to global climate change, *Journal of Phycology* 28:579-590.
Bouma, M.J., and C. Dye
    1997   Cycles of malaria associated with El Niño in Venezuela. *Journal of the American Medical Association* 278:1772-1774.
Bouma, M.J., H.E. Sondorp, and J.H. van der Kaay
    1994a   Health and climate change. *Lancet* 343:302.
    1994b   Climate change and periodic epidemic malaria. Lancet 343:1440.
Brock, J.C., C.R. McClain, M.E. Luthur, and W.W. Hay
    1991   The phytoplankton blooms in the Northwestern Arabian Seas during the Southwest monsoon of 1979. *Journal of Geophysical Research* 96(C 11)20:613-622.
Brock, J.C., and C.R. McClain
    1992   Interannual variability in phytoplankton blooms observed in the Northwestern Arabian Sea during the Southwest monsoon. *Journal of Geophysical Research* 97:733-375.
Burgos, J.J.
    1990   Analogias agroclimatologicas utiles para la adaptacion al posible cambio climatico global de America del Sur. *Revista Geofisica* 32:79-95.
Burgos, J.J., S.I. Curto de Casas, R.U. Carcavallo, and G.I. Galindez
    1994   Global climate change in the distribution of some pathogenic complexes. *Entomologia y Vectores* 1:69-82.
Cabello, F.
    1991   Una visita a un antiguo paradigma en Chile: Deterioro economico y social y epidemias. *Interciencia* 16:176-181.
Centers for Disease Control and Prevention
    1994   Update: Coccidioidomycosis—California, 1991-1993. *MMWR* 43:421-423.
Cockburn, T.A., and J.G. Cassanos
    1960   Epidemiology of epidemic cholera. *Public Health Reports* 75:791.
Colwell, R.R., P.R. Brayton, D.J.Grimes, S.A.Ruszak, H. Hug, and L.M. Palmer
    1985   Viable but non-culturable Vibrio cholerae and related pathogens in the environment: Implications for release of genetically engineered microorganisms, *Biotechnology* 3:817-820.

Dister, S., L. Beck, B. Wood, R. Falco, and D. Fish
  1993   The use of remote sensing technologies in a landscape approach to the study of Lyme
          disease transmission risk. Pp. 1149-1155 in *Proceedings of GIS '93:* Seventh Annual
          Symposium in Geographic Information Systems in Forestry, *Environmental and Natural
          Resource Management.*
Dobson, A., and R. Carper
  1993   Biodiversity. *Lancet* 342:1096-1099.
Edman, J.D., R. Timperi, and D. Werner
  1993   Epidemiology of Eastern Equine Encephalitis in Massachusetts. *Journal of the Florida
          Mosquito Control Association* 64:84-96.
Epstein, P.R., and G.P. Chikwenhere
  1994   Biodiversity questions (letter). *Science* 265:1510-1511.
Epstein, P.R., T.E. Ford, and R.R. Colwell
  1993   Marine ecosystems. *Lancet* 342:1216-1219.
Epstein, P.R., D.J. Rogers, and R. Slooff
  1993   Satellite imaging and vector-borne disease. *Lancet* 341:1404-1406.
Feldman, G., D. Clark, and D. Halpern
  1984   Satellite color observations of the phytoplankton distribution in the Eastern Equatorial
          Pacific during the 1982-1983 El Niño. *Science* 226:1069-1071.
Focks, D.A., E. Daniels, D.G. Haile, and L.E. Keesling
  1995   A simulation model of the epidemiology of urban dengue fever: Literature analysis,
          model development, preliminary validation, and samples of simulation results. *American
          Journal of Tropical Medicine and Hygiene* 53:489-506.
Gill, C.A.
  1920a   The relationship between malaria and rainfall. *Indian Journal of Medical Research*
          37:618-632.
  1920b   The role of meteorology and malaria. *Indian Journal of Medical Research* 8:633-693.
Glantz, M.H., R.W. Katz, and N. Nicholls
  1991   *Teleconnections Linking Worldwide Climate Anomalies. Scientific Basis and Societal
          Impact.* New York: Cambridge University Press.
Graham, N.E.
  1995   Simulation of recent global temperature trends. *Science* 267:666-671.
Hallaegraeff, G.M.
  1993   A review of harmful algal blooms and their apparent global increase. *Phycologia* 32:79-
          99.
Imhoff, M.L., and D.B. Gesch
  1990   The derivation of a sub-canopy digital terrain model of a flooded forest using synthetic
          aperture radar. *Photogrammetric Engineering and Remote Sensing* (11 June):1157-1162.
Imhoff, M.L., and S.W. McCandless
  1988   Flood boundary delineation through clouds and vegetation using L-band space-borne
          radar: A potential new tool for disease vector control programs. *Acta Astranautica*
          17(9):1003-1007.
Kaladis, G.N., and H.F. Diaz
  1989   Global climatic anomalies associated with extremes in the southern oscillation. *Journal of
          Climate* 2:1069-1090.
Martens, W.J.M., J. Rotmans, and L.W. Niessen
  1994   *Climate Change and Malaria Risk: An Integrated Modelling Approach.* GLOBE Report
          Series No. 3 RIVM Report No. 461502003. Bilthoven, The Netherlands: Global Dynam-
          ics and Sustainable Development Programme.

Martens, W.J.M., T.H. Jeffen, and D.A. Focks
  1997   Sensitivity of malaria, schistosomiasis, and dengue to global warming. *Climatic Change* 35:145-156.
Matsuoka, Y., and K. Kai
  1994   An estimation of climatic change effects on malaria. *Journal of Global Environment Engineering* 1:1-15.
Nicholls, N.
  1993   El Niño-Southern Oscillation and vector-borne disease. *Lancet* 342:1284-1285.
Parrilla, G., A. Lavin, H. Bryden, M. Garcia, R. Millard
  1994   Rising temperatures in the sub-tropical North Atlantic Ocean over the past 35 years. *Nature* 369:48-51.
Rogers, D.J., and M.J. Packer
  1993   Vector-borne diseases, models, and global change. *Lancet* 342:1282-1284.
Susskind, J., P. Piraino, L. Rokke, L. Iredell, and A. Mehta
  1997   Characteristics of the TOVS Pathfinder Path A data set. *Bulletin of the American Meteorological Society* 78:1449-1472.
Thompson, L.G., E. Mosley-Thompson, M.E. Davis, et al.
  1995   Late glacial stage and holocene tropical ice core records from Huascarán, Peru. *Science* 269:46-50.
Thwaites, T.
  1994   Are the antipodes in hot water? *New Scientist* 12(November):21.
Walker, B.H.
  1994   Landscape to regional-scale responses of terrestrial ecosystems to global change. *Ambio* 23:67-73.

ADDITIONAL READINGS

Anon
  1994   Indian Ocean may have El Niño of its own. *EOS Transactions*, American Geophysical Union. 75(December 13):585-586.
Bengtsson, L., U. Schlese, E. Roeckner, M. Latif, T.P. Barnett, and N. Graham
  1993   A two-tiered approach to long-range climate forecasting. *Science* 261:1026-1029.
Bindoff, N.L., and J.A. Church
  1992   Warming of the water column in the southwest Pacific. *Nature* 357:59-62.
Broecker, W.S.
  1987   Unpleasant surprises in the greenhouse? *Nature* 328:123-126.
Chang, P., J.I. Link, and L.I. Hong
  1997   A decadal climate variation in the tropical Atlantic Ocean from thermodynamic air-sea interactions. *Nature* 385:516-518.
Dansgaard, W., S.J. Johnson, H.B. Clausen, et. al.
  1993   Evidence for general instability of past climate from a 250-yr ice-core record. *Nature* 364:218-220.
Darzi, M., J.K. Firestone, G. and C.R. McClain
  1991   Current efforts regarding the SEAPAK oceanographic analysis system. Pp. 109-115 in *Proceedings of 7th Conference on Interactive and Informative Processing Systems for Meteorology, Hydrology, and Oceanography.* Boston, Mass.: American Meteorological Society.
Easterling, D.R., B. Horton, P. D. Jones, T.C. Peterson, T.R. Karl, D.E. Parker, M.J. Salinger, V. Razuvayev, N. Plummer, P. Jamason, and C.K. Folland
  1997   Maximum and minimum temperature trends for the globe. *Science* 277:363-367.

Epstein, P.R.
    1993   Algal blooms in the spread and persistence of cholera. *BioSystems* 31:209-221.
Epstein, P.R., O.C. Pena, and J.B. Racedo
    1995   Climate and disease in Colombia. *Lancet* 346:1243-1244.
Feldman, G., et. al.
    1989   Ocean color: Availability of the global data set, *EOS Transactions*, American Geophysical Union 70 (23):634-635, 645-641.
Gordon, H.R., D.K. Clark, J.W. Brown, O.B. Brown, R.H. Evans, and W.W. Bronkow
    1983   Phytoplankton pigment concentration in the Middle Atlantic Bight: Comparison of ship determinations and CZCS estimates, *Applied Optics* 22 (1):20-36.
Graham, N.E.
    1995   Simulation of recent temperature trends. *Science* 267:666-671.
Greenland Ice-Core Project (GRIP)
    1993   Climate instability during the last interglacial period recorded in the GRIP ice core. *Nature* 364:203-207.
Holligan, P.M., M. Viollier, D.S. Harbour, P. Kamus, and N. Champaque, Philippe
    1983   Satellite and ship studies of coccolithophore production along our continental shelf edge. *Nature* 304(28):339-342.
Holliday, D.V.
    1993   Applications of advanced acoustic technology in large marine ecosystem studies. Pp. 301-319 in *Large Marine Ecosystems: Stress, Mitigation, and Sustainability*, K. Sherman, L.M. Alexander, B. D. Gold, eds. Washington, D.C.: AAAS Press.
Karl, T.R., P.D. Jones, R.W. Knight, G. Kukla, N. Plummer, V. Razuvayev, K.P. Gallo, J. Lindsay, R.J. Charlson, and T.C. Peterson
    1993   A new perspective on recent global warming: Asymmetric trends of daily maximum and minimum temperature. *Bulletin of the American Meteorological Soc*iety 74:1007-1023.
Karl, T.R., R.W. Knight, D.R. Easterling, and R.G. Quayle
    1995a   Trends in U.S. climate during the twentieth century. *Consequences* 1:3-12.
Karl, T.R., R.W. Knight, and N. Plummer
    1995b   Trends in high-frequency climate variability in the twentieth century. *Nature* 377:217-220.
Karl, T.R., N. Nicholls, and J. Gregory
    1997   The coming climate. *Scientific American* (May):78-83.
Kiladis, G.N., and H.F. Diaz
    1989   Global climatic anomalies associated with extremes in the Southern Oscillation. *Journal of Climate* 2:1069.
Loevinsohn, M.
    1994   Climatic warming and increased malaria incidence in Rwanda. *Lancet* 343:714-718.
Manabe, S., and R.J. Stouffer
    1993   Century-scale effects of increased atmospheric CO_2 on the ocean-atmosphere system. *Nature* 364:215-218.
Martens, W.J.M.
    1995   *Modelling the Effect of Global Warming on the Prevalence of Schistosomiasis.* Report No. 461502010. Bilthoven, The Netherlands: National Institute of Public Health and the Environment (RIVM).
Martin, P.H., and M.G. Lefebvre
    1995   Malaria and climate: Sensitivity of malaria potential transmission to climate. *Ambio* 24:200-207.
Meehl, G.A., and W.M. Washington
    1993   South Asian summer monsoon variability in a model with double atmospheric carbon dioxide concentration. *Science* 260:1101-1104.

Mayewski, P.A., L.D. Meeker, S. Whitlow, et al.
  1993   The atmosphere during the Younger Dryas. *Science* 261:195-197.
Parrilla, G., A. Lavin, H. Bryden, M. Garcia, and R. Millard
  1994   Rising temperatures in the sub-tropical North Atlantic Ocean over the past 35 years.
          Nature 369:48-51.
Pearce, F.
  1997   Southern oceans hold key to climate. *New Scientist* 5(April):21.
Powers, D.A.
  1993   Application of molecular techniques to large marine ecosystems. Pp. 320-352 in: *Large
          Marine Ecosystems: Stress, Mitigation, and Sustainability*, K. Sherman, L.M. Alexander,
          B. D. Gold, eds., Washington, D.C.: AAAS Press.
Regaldo, A.
  1995   Listen up! The world's oceans may be starting to warm. *Science* 268: 1436-1437.
Travis, J.
  1994   Taking a bottom-to sky "slice" of the Arctic Ocean. *Science* 266:1947-1948.
Trenberth, K.E., and T.J. Hoar
  1996   The 1990-1995 El Niño-Southern Oscillation event: Longest on record. *Geophysical
          Research Letters* 23:57-60.
Tziperman, E.
  1997   Inherently unstable climate behavior due to weak thermohaline ocean circulation. *Nature*
          386:592-595.
Yoder, J.A., and G. Garcia-Moliner
  1994   Application of satellite remote sensing and optical buoys/moorings to LME studies.
          Ambio 23:353-358.
Zebiak, S.E., and M.A. Cane
  1991   Natural climate variability in a coupled model. In *Greenhouse Gas-Induced Climatic
          Change: A Critical Appraisal of Simulations and Observations*, M.E. Schlesinger, ed.
          Amsterdam, The Netherlands: Elsevier Science Publishers BV.

Appendix A

An Annotated Guide to Earth Remote Sensing Data and Information Resources for Social Science Applications

Robert S. Chen

The volume and diversity of earth remote sensing data and the range of applications to social science research are growing rapidly. This annotated guide presents a selected set of information resources and entry points for those interested in exploring and using earth remote sensing data in interdisciplinary research, applications, and education. It focuses primarily on electronic data and information resources available via the Internet, since these tend to have the most up-to-date information on availability and access.

The resources listed are organized into six categories, each with the subcategories shown:

I. Basic Information and Documentation on Earth Remote Sensing
    I.A   General Overviews and Compendia
    I.B   Frequently Asked Questions (FAQ)
    I.C   Glossaries
    I.D   Tutorials and Other Educational Materials
II. Selected Listings of Earth Remote Sensing Data and Information Resources
    II.A   Comprehensive Lists
    II.B   Selective Lists
III. Selected Remote Sensing Data Catalogs and Search Tools
    III.A   Multiple Data Center Catalogs and Search Tools
    III.B   Selected Data Center Catalogs and Search Tools

IV. Selected Earth Remote Sensing Data Systems and Centers
   IV.A   Selected Mutilational Data Systems and Centers
   IV.B   Selected National Centers
V. Other Noteworthy Resources Related to Earth Remote Sensing
   Applications and Education
VI. Selected E-Mail/WWW Gateways

Within each subcategory, items are listed alphabetically. Suggested starting points for general readers are indicated with an asterisk.

It should be noted that the inclusion of commercial data and information resources is not intended as an endorsement of any products or services provided. Moreover, given the rapidly expanding wealth of data and information resources, not all interesting and useful sites could be included in this guide. However, a sampling of noteworthy sites of likely interest to social scientists is provided to give potential users a flavor of what is now becoming available. This guide does not attempt to cover the wide range of sites related to geographic information systems or social science data, although many of the sites included here do contain information relevant to these subjects.

Both the content and addresses of Internet sites tend to change over time. The annotations and Internet locations given here reflect the status of sites circa September 1997. For an up-to-date, hyperlinked version of this appendix, visit the Web site of the Socioeconomic Data and Applications Center (SEDAC) at the Uniform Resource Locator (URL):<*http://sedac.ciesin.org/remote*>. Users are also encouraged to utilize the lists and search tools included here to obtain current information on existing and new resources.

Users who do not have direct WWW access to the Internet may be able to access many of the data and information resources listed through an electronic mail (e-mail) WWW gateway. Section VI lists data services of this type.

I. BASIC INFORMATION AND DOCUMENTATION ON EARTH REMOTE SENSING

I.A General Overviews and Compendia

*I.A-1 **Atlas of Satellite Observations Related to Global Change.** Published in 1993, this volume includes more than 25 articles on the use of satellite data and imagery to address key scientific questions concerning global environmental change. Topics covered include the earth's radiation balance, atmosphere, oceans, land, cryosphere, and the effects of human activities. An appendix describes both existing and planned operational and research satellites for earth observation, by country. AVAILABILITY: Printed volume available from Cambridge University Press (ISBN 0-521-43467-X).

I.A-2 **Earth Observations from Space: History, Promise, and Reality.**
The Space Studies Board of the U.S. National Research Council has prepared a
detailed history of the Mission to Planet Earth Program and reviewed the overall
development and organization of U.S. civil earth observation programs. Published in 1995, the report provides useful programmatic background for understanding the current and planned suite of earth remote sensing satellites and
associated instruments. AVAILABILITY: Printed version available from the Space
Studies Board, National Research Council, 2101 Constitution Ave., N.W., Washington, D.C. 20418. Executive Summary available on line via the National
Academy Press Reading Room at URL *<http://www.nap.edu/readingroom>*
(search on title). A direct link to the summary is at URL *<http://www.nap.edu/
readingroom/records/NX006300.html>*.

I.A-3 **Manual of Photographic Interpretation, 2nd Edition.** Updated in
1997 from the original published in 1960, this comprehensive manual reviews the
history and fundamentals of photographic interpretation; methods for recogniz-
ing and assessing landforms, soils, vegetation, hydrology, and structures and
cultural features; and a range of applications including land-use and land-cover
inventory, agriculture and forestry, wildlife and wetlands, urban planning and
archaeology, military uses, and environmental monitoring. It includes two appendices that list major imagery collections and sources of aerial photographs.
AVAILABILITY: Printed volume available from the American Society for Photogrammetry and Remote Sensing, 5420 Grosvenor Lane, Suite 210, Bethesda,
Maryland 20814-2160 (ISBN 1-57083-039-8). For additional information see
URL *<http://www.us.net/asprs/publications/bookstore/manuals.html>*.

I.A-4 **Manual of Remote Sensing, 3rd Edition (a series): Earth Observing Platforms and Sensors.** Now published in Windows-compatible CD-ROM
format, this manual, updated in 1996, provides detailed information on more than
270 satellite and airborne platforms focusing on atmospheric, geophysical, and
earth resources. The CD-ROM includes sample data sets, technical specifications of hardware, and numerous color illustrations. AVAILABILITY: CD-ROM
available from the American Society for Photogrammetry and Remote Sensing,
5420 Grosvenor Lane, Suite 210, Bethesda, Maryland 20814-2160. For additional information see URL *<http://www.us.net/asprs/publications/bookstore/
manuals.html>*.

I.A-5 **MTPE 1995 EOS Reference Handbook.** The Mission to Planet
Earth (MTPE) Program of the U.S. National Aeronautics and Space Administra-
tion (NASA) maintains a reference handbook describing the status of the Earth
Observing System (EOS). This document includes detailed information on
planned EOS and related satellite missions, EOS instruments, EOS interdiscipli-
nary science investigations, and MTPE pathfinder data sets (see item V-17).
AVAILABILITY: URL *<http://eospso.gsfc.nasa.gov/eos_reference/TOC.html>*.
Printed version available from EOS Project Science Office, Code 900, NASA/
Goddard Space Flight Center, Greenbelt, Maryland 20771, tel. (301) 286-3411.

*I.A-6 **Thematic Guide on the Use of Satellite Remote Sensing To Study the Human Dimensions of Global Environmental Change.** This on-line guide, developed by the Consortium for International Earth Science Information Network (CIESIN), provides an overview of satellite remote sensing in relationship to research on human interactions with global environmental change. It includes the full text of selected literature on remote sensing applications, as well as other bibliographic materials and direct links. AVAILABILITY: URL *<http://www. ciesin.org/>*.

*I.A-7 **Trends in Commercial Space—1996: Satellite Remote Sensing.** This paper, prepared by the Office of Air and Space Commercialization of the U.S. Department of Commerce as part of an edited volume, provides an overview of commercial satellite remote sensing. It summarizes the Presidential Policy on Commercial Remote Sensing and lists planned worldwide commercial remote sensing capabilities. AVAILABILITY: URL *<http://cher.eda.doc.gov/oasc/rmtsens.html>*.

I.B Frequently Asked Questions (FAQ)

I.B-1 **A FAQ on Vegetation in Remote Sensing.** This FAQ document addresses the use of remote sensing to assess vegetation characteristics. It includes details on the derivation of standard indices and a bibliography. The on-line version is dated 13 October 1994. AVAILABILITY: URL *<http://www.info-mine.com/technomine/ege/vegfaq.txt>*.

*I.B-2 **The Satellite Imagery FAQ.** This five-part FAQ provides summary information on satellite-based imagery of the earth. It deals with both general and technical questions and includes an extensive list of remote sensing satellites and instruments, with pointers to resources available on the Internet. The FAQ is updated approximately monthly by members of the Image Processing and Remote Sensing Listserver (IMAGRS-L). (See also item II.A-9.) AVAILABILITY: URL *<http://www.geog.nott.ac.uk/remote/satfaq.html>* (latest version).

I.C Glossaries

I.C-1 **CCRS Remote Sensing Glossary.** This searchable database, maintained by the Canada Centre for Remote Sensing, contains a comprehensive list of terms related to radar, optical, and airborne remote sensing and a dictionary of acronyms. Results may be displayed in English or French. (See also items III.B-1 and IV. B-2.) AVAILABILITY: URL *<http://www.ccrs.nrcan.gc.ca/ccrs/eduref/ ref/glosndxe.html>*.

I.C-2 **Dictionary of Abbreviations and Acronyms in Geographic Information Systems, Cartography, and Remote Sensing.** An extensive list of abbreviations and acronyms related to geographic information systems (GIS) and remote sensing is provided (dated October 1997). AVAILABILITY: URL *<http:// www.lib.berkeley.edu/EART/abbrev.html>*.

I.C-3 **Earth Observing System (EOS) Acronyms/Abbreviations.** A comprehensive list of acronyms and abbreviations is provided by the EOS Project Science Office at NASA's Goddard Space Flight Center. AVAILABILITY: URL *<http://eospso.gsfc.nasa.gov/EOS_acronym.html>*.

I.C-4 **Earth Observing System (EOS) Glossary.** This is an extensive glossary of concepts and terms related to remote sensing and weather, aimed primarily at teachers and students in Grades 7-12. A short bibliography and a list of educational resources around the United States are also provided. AVAILABILITY: URL *<http://eospso.gsfc.nasa.gov/Earth_Glossary/Intro.html>*.

I.C-5 **Global Land Information System (GLIS) Glossary.** An extensive glossary of earth science acronyms and terms is available from the U.S. Geological Survey's GLIS, maintained by the Earth Resources Observation System (EROS) Data Center (EDC) (see item III.B-4). AVAILABILITY: URL *<http://edcwww.cr.usgs.gov/glis/hyper/glossary/index>*.

I.C-6 **Glossary of Remote Sensing Terminology and Acronyms.** A brief list of remote sensing terms and acronyms and associated definitions is provided. AVAILABILITY: URL *<http://www.seaspace.com/glossary.html>*.

I.D Tutorials and Other Educational Materials

*I.D-1 **Community Colleges for Innovative Technology Transfer, Inc. (CCITT) Project in Remote Sensing, Image Processing and Geographic Information Systems.** CCITT is a consortium of U.S. community colleges affiliated with a local NASA Center. This project is developing a set of curriculum modules aimed at undergraduate courses in earth system sciences or related disciplines. The project also offers periodic faculty workshops in remote sensing, image processing, and GIS. AVAILABILITY: URL *<http://earth.fhda.edu/>*.

I.D-2 **GLOBE Visualization Server.** The visualization server of the Global Learning and Observations to Benefit the Environment (GLOBE) program provides images of global environmental data, including both student measurements and reference data from the Environmental Modeling Center (EMC) and selected satellites. AVAILABILITY: URL *<http://glom.arc.nasa.gov/>*.

*I.D-3 **NASA's Observatorium.** This is a public-access site for earth and space data, oriented primarily to precollege students. It includes an image gallery and selected exhibits on topics such as ozone and the use of remote sensing in agriculture and in the Mississippi flood of 1993. AVAILABILITY: URL *<http://www.rspac.ivv.nasa.gov/nasa/core.shtml>*.

*I.D-4 **Remote Sensing Core Curriculum.** Under development by a distributed team involving the University of Maryland, Baltimore County (UMBC), the National Center for Geographic Information and Analysis (NCGIA), and the University of South Carolina, this curriculum provides an introduction to photo interpretation and photogrammetry, remote sensing of the environment, digital image processing, and remote sensing applications. It includes links to educational and on-line data resources and provides both lecture and exercise materials on topics such as population estimation, environmental warfare, flood analysis, and rangeland management. AVAILABILITY: URL *<http://www.umbc.edu/rscc>* or *<http://research.umbc.edu/~tbenja1/index.html>*.

*I.D-5 **Remote Sensing Task Force.** The Remote Sensing Task Force of the National Council for Geographic Education (NCGE) has developed a set of instructional materials dealing with remote sensing of the environment. Specific topics include the Persian Gulf War, the nuclear accident at Chernobyl, fires in Yellowstone National Park, and the 1993 flood around St. Louis, Missouri. The Web site includes ordering information and samples of low- and high-resolution imagery. AVAILABILITY: URL *<http://www.oneonta.edu/~baumanpr/ncge/rstf.htm>*.

I.D-6 **Remote Sensing Tutorial.** This on-line tutorial under development by Goddard Space Flight Center focuses on the role of space technology in monitoring the earth's surface and atmosphere. It is aimed at both professional audiences and the general public and includes extensive materials on remote sensing applications. Although it is not yet complete, more than half of the planned chapters are available, and several more are apparently in progress. AVAILABILITY: URL *<http://code935.gsfc.nasa.gov/Tutorial/TofC/Coverpage.html>*.

I.D-7 **Satellite Meteorology Online Remote Sensing Guide.** This instructional module focuses on weather satellites, particularly Geostationary Operational Environmental Satellites (GOES) and Polar Orbiting Environmental Satellites (POES). AVAILABILITY: URL *<http://ww2010.atmos.uiuc.edu/(GH)/guides/rs/sat/home.rxml>*.

I-D-8 **Satellite Training Application.** This computer-based training application, under development by the Naval Research Laboratory in Monterey, California, presently includes a prototype satellite tutorial on interpretation of satellite imagery and training materials on the Special Sensor Microwave Imager (SSM/I) of the Defense Meteorological Satellite Program (DMSP). Additional materials are being prepared for other U.S. environmental satellites. AVAILABILITY: URL *<http://www.nrlmry.navy.mil/~white/>*.

II. SELECTED LISTINGS OF EARTH REMOTE SENSING DATA AND INFORMATION RESOURCES

II.A Comprehensive Lists

II.A-1 **Commercial Remote Sensing Program (CRSP) Remote Sensing/GIS Resource List.** Developed by the CRSP Office at NASA's John C. Stennis Space Center, this directory of Web sites covers a range of remote sensing topics, including satellite and airborne remote sensing; radar, lidar, and hyper-spectral imaging; image processing and GIS software; and remote sensing and GIS in education. It includes information on both commercial and noncommercial sources. AVAILABILITY: URL <*http://crsphome.ssc.nasa.gov/TUTORIAL/COMPEND/NEWFPG.HTM*>.

II.A-2 **EWSE: Other Sites Search.** This service generates an alphabetical list of Web sites related to earth observation. Users may then retrieve a short description of each Web site and link directly to that site. For a more specific search, see item III.A-5, the European Wide Service Exchange (EWSE). AVAILABILITY: URL <*http://ewse.ceo.org/anonymous/all/list.pl/1920*>.

II.A-3 **GeoData Information Sources.** The Center for Global and Regional Environmental Research (CGRER) at the University of Iowa maintains a list of hypertext links to digital cartographic and environmental data, including remotely sensed data. Subcategories include remote sensing data documentation, public domain and commercial sources, and meteorological satellite imagery. AVAILABILITY: URL <*http://www.cgrer.uiowa.edu/servers/servers_geodata.html*>.

*II.A-4 **The Journalists' Guide to Remote Sensing Resources.** Maintained by the American University School of Communication, this Web site provides extensive, annotated lists of remote sensing data and information resources. It includes a list of earth imagery resources and contact information for more than 50 countries and a large set of links to U.S. government remote sensing and mapping resources. AVAILABILITY: URL <*http://gurukul.ucc.american.edu/earthnews*>.

*II.A-5 **Remote Sensing Data and Information Directory of Resources.** The Public Use of Remote Sensing Data (RSD) Program at NASA's Goddard Space Flight Center maintains a directory of resources covering weather and climate, space and astronomy, earth images from space, data archives, specific instruments, and projects and organizations. AVAILABILITY: URL <*http://rsd.gsfc.nasa.gov/rsd/RemoteSensing.html*>.

II.A-6 **Remote Sensing Divisions: Recommended Links.** The Department of Geography at the University of Zurich in Switzerland provides an extensive set of recommended links, focusing in particular on European resources and covering organizations and institutes, data archives and services, satellites and instruments, navigation, press releases, and other resources. AVAILABILITY: URL <*http://www.geo.unizh.ch/rsl1/rs_links/*>.

II.A-7 **Remote Sensing Guide.** This well-organized listing is maintained at Brunel University in the United Kingdom. It includes direct links to major remote sensing lists and indexes, satellite and sensor sites, organizational information, publications, educational resources, software, and mailing lists and news groups. AVAILABILITY: URL <*http://http2.brunel.ac.uk:8080/depts/geo/Contents.html*>.

II.A-8 **Remote Sensing Resources on the Web.** This comprehensive listing of remote sensing resources covers satellites and sensors; remote sensing research programs; radar, aerial, and extraterrestrial remote sensing; and publications, software, and other resources. AVAILABILITY: URL <*http://midas.ac.uk/rs/*>.

*II.A-9 **The Satellite Imagery FAQ, Parts 4 and 5.** These sections of the FAQ (item I.B-2) provide a detailed inventory of remote sensing data sources, including lists of sites by satellite platform and country and a list of commercial sites. It is updated approximately monthly by members of the Image Processing and Remote Sensing Listserver (IMAGRS-L). AVAILABILITY: URL *<http:// www.geog.nott.ac.uk/remote/satfaq.html>* (latest version).

II.A-10 **Some Interesting WWW Pages.** Developed as a service for members of The Remote Sensing Society in the United Kingdom, this page includes numerous links related to data, sensors, satellite platforms, publications, conferences, and other topics. AVAILABILITY: URL *<http://www.geog.nottingham.ac.uk/ ~mather/useful_links.html>*.

II.A-11 **Space Mission Acronym List and Hyperlink Guide.** This on-line guide, developed by NASA's Instrument and Sensing Technology Program, provides information on a wide range of space missions, including earth observation satellites and direct links to relevant platform and instrument information. AVAILABILITY: URL *<http://ranier.hq.nasa.gov/Sensors_page/MissionLinks.html>*.

II.A-12 **WWW Virtual Library: Remote Sensing.** Developed by the Technical Research Centre of Finland, this directory provides a long list of links to remote sensing resources and related organizations around the world. Organizational lists are divided by region; other topics include data sources, societies, conferences, documents, journals, news groups, mailing lists, and related fields. AVAILABILITY: URL *<http://www.vtt.fi/aut/rs/virtual/>*.

II.A-13 **Yahoo! Science/Geology_and_Geophysics/Remote_Sensing.** This browse page provided by Yahoo lists a range of links relevant to remote sensing. Subcategories include conferences and institutes; cross-links are provided to companies and earth pictures. AVAILABILITY: URL *<http://www.yahoo. com/yahoo/Science/Geology_and_Geophysics/Remote_Sensing/>*.

II. B Selective Lists

II.B-1 **Africa Data Dissemination Service (ADDS).** Developed by the Earth Resources Observation System (EROS) Data Center (see item IV.B-4), this service lists continental, regional, and national data for Africa, including holdings of Normalized Difference Vegetation Index (NDVI) data based on the Advanced Very High Resolution Radiometer (AVHRR). AVAILABILITY: URL *<http:// edcintl.cr.usgs.gov/adds/>*.

II.B-2 **CHAART Data Sources on the Internet.** The Center for Health Applications of Aerospace Related Technologies (CHAART) at NASA's Ames Research Center has developed an annotated list of remote sensing and GIS data sites. The focus is primarily on federal data resources. AVAILABILITY: URL *<http://geo.arc.nasa.gov/sge/health/links/links.html>*.

II.B-3 **Imaging Intelligence on the Web.** Developed by the Intelligence Reform Project of the Federation of American Scientists, this site focuses on imagery intelligence resources on the WWW. It includes an image gallery and links to key intelligence documents and archives. AVAILABILITY: URL <*http:// www.fas.org/irp/wwwimint.html*>.

II.B-4 **Japan GIS/Mapping Sciences Resource Guide.** This is an on-line guide linking to an array of remote sensing, mapping, and imagery resources related to Japan, including distribution centers, software, organizations, and institutes. (see also items III.B-3, IV.B-8, and IV.B-9.) AVAILABILITY: URL <*http:// www.cast.uark.edu/jpgis/*>.

III. SELECTED REMOTE SENSING DATA CATALOGS AND SEARCH TOOLS

III.A Multiple Data Center Catalogs and Search Tools

*III.A-1 **The CIESIN Gateway.** The CIESIN Gateway, developed by the Consortium for International Earth Science Information Network (CIESIN), provides integrated searching of multiple catalogs that conform to the Z39.50 international standard, including the Global Change Master Directory (GCMD), various National Oceanic and Atmospheric Administration (NOAA) catalogs, and CIESIN's own metadata holdings. The Gateway supports both full-text search and Boolean searches on specified fields. AVAILABILITY: URL <*http://wwwgateway.ciesin.org*>.

III.A-2 **Department of Defense Master Environmental Library (MEL).** MEL is designed to provide a single point for searching U.S. Department of Defense environmental data. It provides access to metadata developed by distributed regional sites. Users may specify a location, time period, keyword(s), and database of interest. Both Java and non-Java query interfaces are currently provided. AVAILABILITY: URL <*http://www-mel.nrlmry.navy.mil/homepage.html*>.

*III.A-3 **Earth Pages.** Earth Pages is an on-line search and index tool for earth science data, information, and resources on the Internet. Resources are organized by topic or alphabetically or may be searched using keywords. Organizations must register in order to be included in the database. AVAILABILITY: URL <*http://starsky.hitc.com/earth/earth.html*>.

III.A-4 **EOSDIS Information Management System (IMS).** The IMS provides access to the extensive data holdings of NASA's Earth Observing System Data and Information System (EOSDIS). A graphical user interface (GUI) based on the X Window System™ supports data searching and browsing, access to directory and guide information, and data ordering. A WWW-based interface provides more limited search and access capabilities. AVAILABILITY: For the WWW interface, see URL <*http://harp.gsfc.nasa.gov/v0ims/*>. For the GUI, see URL <*http://www-v0ims.gsfc.nasa.gov/v0ims/ims_access.html*> for a list of telnet access points.

III.A-5 **European Wide Service Exchange (EWSE).** The EWSE is an on-line information exchange testbed of the Centre for Earth Observation (CEO) Programme, designed to help the earth observation community find relevant data and resources (see also item IV.A-1). The service supports free text searching, geographic and keyword searching, and browsing of a dynamic database. EWSE also provides access to gateways compatible with the NASA Information Management System (see item III.A-4). AVAILABILITY: URL *<http://ewse.ceo.org/>*.

*III.A-6 **The Global Change Master Directory (GCMD).** The GCMD provides access to an extensive set of metadata records describing remote sensing, in situ, and other earth science data. The GCMD search interface permits both free-text and controlled searches (using standard search terms). The GCMD is the American Coordinating Node of the Committee on Earth Observation Satellites International Directory Network (CEOS IDN). AVAILABILITY: URL *<http://gcmd.gsfc.nasa.gov/>*.

III.A-7 **Global Environmental Information Locator Service (GELOS).** GELOS is an alpha version search tool developed as part of the G7 Environment and Natural Resources Management project (G7-ENRM). The G7-ENRM project seeks to create a globally distributed virtual library of environmental and natural resource data. The current version of GELOS supports free-text, geographical, and keyword searches. AVAILABILITY: URL *<http://enrm.ceo.org/>*.

III.A-8 **NOAA Environmental Services Data Directory.** This catalog includes metadata for the extensive environmental data holdings of the U.S. National Oceanic and Atmospheric Administration (NOAA). It includes the holdings of many NOAA data centers, such as the Satellite Active Archive (SAA), the National Climatic Data Center (NCDC), the National Geophysical Data Center (NGDC), and the National Oceanographic Data Center (NODC). The catalog supports both full-text search and custom searches utilizing fields corresponding to the Governmental Information Locator System (GILS) or the Federal Geographic Data Committee (FGDC) Content Standards for Digital Geospatial Metadata. (See also item IV.B-10.) AVAILABILITY: URL *<http://www.esdim. noaa.gov/NOAA-Catalog/>*.

III.B Selected Data Center Catalogs and Search Tools

III.B-1 **CCRS Earth Observation Catalogue (CEOCat).** The Canada Centre for Remote Sensing (see items I.C-1 and IV.B-2) maintains an on-line inventory of "quick look" images for its holdings of Landsat, Advanced Very High Resolution Radiometer (AVHRR), Système pour l'Observation de la Terre (SPOT), and synthetic aperture radar (SAR) data. AVAILABILITY: URL *<http:// www.ccrs.nrcan.gc.ca/ccrs/>*.

III.B-2 **CRISP Catalogue Browse.** This resource permits browsing of Système pour l'Observation de la Terre (SPOT), European Remote Sensing Satellite (ERS), and RADARSAT imagery for South and Southeast Asia available from the Center for Remote Imaging, Sensing and Processing (CRISP), the National University of Singapore. AVAILABILITY: URL *<http://www.crisp.nus.sg/>*.

III.B-3 **EOIS WWW Image Catalogue Service.** This prototype interface provides inventory-level search for the holdings of the Earth Observation Information System (EOIS) of the Japanese National Space Development Agency (NASDA). Searches may be conducted based on area, satellite, sensor, date, and cloud cover for the Japan area or the globe. Both English and Japanese versions of the interface are available. (See also items II.B-4, IV.B-8, and IV.B-9.) AVAILABILITY: URL *<http://hgssac01.eoc.nasda.go.jp/~goin/index.html>*.

III.B-4 **Global Land Information System (GLIS).** The U.S. Geological Survey's GLIS provides on-line access to the extensive libraries of land-oriented remote sensing imagery held by the U.S. government and Canadian and European sources (see also item I.C-5). Three different interfaces are supported: an X Window™ client, a PC client, and a WWW interface (including both "frames" and "nonframes" versions). The client interfaces support more interactive and powerful search and browse functions. AVAILABILITY: URL *<http://edcwww.cr.usgs.gov/glis/glis.html>* for WWW access and for information on downloading clients.

IV. SELECTED EARTH REMOTE SENSING DATA SYSTEMS AND CENTERS

IV.A Selected Multinational Data Systems and Centers

IV.A-1 **Centre for Earth Observation (CEO).** The CEO Programme is a European Community program for advancing the use of earth observation data. The WWW site provides access to a set of demonstration case studies related to satellite orbit tracking, geographic information and visualization, and selected applications. CEO has developed the European Wide Service Exchange (EWSE) testbed (see item III.A-5) for the exchange of earth observation data. AVAILABILITY: URL *<http://ceo-www.jrc.it/>*.

*IV.A-2 **Earthnet** *online.* The European Space Agency (ESA) maintains access to its satellite data resources via the earthnet *online* service, including data from both European Remote Sensing Satellites (ERS) and non-ESA satellites. Instruments include microwave sounder, synthetic aperture radar, altimeter, Along Track Scanning Radiometer (ATSR), and wind scatterometer. AVAILABILITY: URL *<http://gds.esrin.esa.it/>*.

IV.A-3 **The World Data Center System.** Many of the World Data Centers (WDCs) of the International Council of Scientific Unions (ICSU) hold remote sensing data of various types. There are 13 discipline centers in the United States (WDC-A), 7 in Russia (WDC-B), 9 in Europe (WDC-C1), 7 in Japan and India (WDC-C2), and 9 in China (WDC-D). This Web site contains a guide to the WDC system and links to WDCs that maintain their own home pages. AVAILABILITY: URL *<http://www.ngdc.noaa.gov/wdc/wdcmain.html>*.

IV.B Selected National Centers

IV-B-1 **Australian Centre for Remote Sensing (ACRES).** ACRES is Australia's major satellite remote sensing organization, an element of the Australian Surveying and Land Information Group (AUSLIG), Department of Administrative Services. ACRES provides access to satellite data from Landsat, Système pour l'Observation de la Terre (SPOT), European Remote Sensing Satellites (ERS), RADARSAT, Japanese Remote Sensing Satellites (JERS), and other platforms. The Web site includes a digital catalogue of available data. AVAILABILITY: URL <*http://www.auslig.gov.au/acres/index.htm*>.

IV.B-2 **Canada Centre for Remote Sensing (CCRS).** The CCRS oversees the acquisition of earth observation data and coordinates remote sensing applications in Canada. It provides access to the CCRS Earth Observation Catalogue (CEOCat) image archive (see item III.B-1), including extensive Système pour l'Observation de la Terre (SPOT), Landsat, and RADARSAT imagery, and maintains a database of remote sensing companies in Canada. AVAILABILITY: URL <*http://www.ccrs.nrcan.gc.ca/ccrs/*>.

IV.B-3 **Global Hydrology Research Center (GHRC).** Formerly the Marshall Distributed Active Archive Center (DAAC), the GHRC continues to archive and disseminate hydrological data, including a range of cloud, water, and lightning data. The site provides access to an on-line search tool, Hydro, and to selected data sets via File Transfer Protocol (FTP). AVAILABILITY: URL <*http://ghrc.msfc.nasa.gov/*>. Also see URL <*http://ghrc.msfc.nasa.gov/uso/MSFC.DAAC.Data.Trans.html*> for information about the transfer of data from the former Marshall DAAC to other DAACs (see item IV.B-4).

*IV.B-4 **NASA Distributed Active Archive Centers (DAACs).** Eight NASA DAACs provide data management and dissemination support for a wide range of existing NASA remote sensing data platforms and upcoming Earth Observing System (EOS) missions:

• **Alaska Synthetic Aperture Radar (SAR) Facility.** Includes SAR data from the European Remote Sensing Satellite (ERS)-1, the Japanese Earth Remote Sensing Satellite (JERS)-1, and the Canadian RADARSAT mission, plus selected Advanced Very High Resolution Radiometer (AVHRR), Landsat, and high-altitude aerial imagery. URL <*http://www.asf.alaska.edu/*>.

• **Earth Resources Observation System (EROS) Data Center.** Archives data on land processes from Spaceborne Imaging Radar-C (SIR-C), Landsat pathfinders (see item V-17), global 1-km AVHRR, and selected airborne platforms. URL <*http://edcwww.cr.usgs.gov/landdaac/*>. Maps, photographs, and special project images are also available from the EROS home page at URL <*http://edcwww.cr.usgs.gov/eros-home.html*>.

• **Goddard Space Flight Center (GSFC) DAAC:** Is responsible for data on upper atmosphere, global biosphere, and atmospheric dynamics, including data from the Upper Atmosphere Research Satellite (UARS), Television InfraRed

Observation Satellite (TIROS) Operational Vertical Sounder (TOVS), Sea-Viewing Wide-Field-of-View Sensor (SeaWiFS), Nimbus-7 Coastal Zone Color Scanner (CZCS), and Tropical Ocean Global Atmosphere-Coupled Ocean Atmosphere Response Experiment (TOGA-COARE). URL *<http://daac.gsfc.nasa.gov/>*.

• **Jet Propulsion Laboratory Physical Oceanography (JPL PO) DAAC.** Manages physical oceanography data, including data from the Ocean Topography Experiment (TOPEX/Poseidon), National Oceanic and Atmospheric Administration (NOAA) Advanced Very High Resolution Radiometer (AVHRR), Nimbus 7, and Seasat. Future holdings will include scatterometer and altimeter data from multiple platforms. URL *<http://podaac-www.jpl.nasa.gov/>*.

• **Langley Research Center (LaRC) DAAC.** Is responsible for data related to the earth's radiation budget, tropospheric chemistry, clouds, and aerosols, including remote sensing data from the Earth Radiation Budget Experiment (ERBE), International Satellite Cloud Climatology Project (ISCCP), Stratospheric Aerosol and Gas Experiment (SAGE), Stratospheric Aerosol Measurement II (SAM II), Global Tropospheric Experiment (GTE), First ISCCP Regional Experiment (FIRE), and Space Shuttle. URL *<http://eosweb.larc.nasa.gov/>*.

• **National Snow and Ice Data Center (NSIDC).** Oversees data related to snow and ice, cryosphere, and climate, including passive microwave-derived products, Advanced Very High Resolution Radiometer (AVHRR) scenes, altimeter data, and field experiment data. URL *<http://www-nsidc.colorado.edu>*.

• **Oak Ridge National Laboratory (ORNL) DAAC.** Archives and distributes data related to biogeochemical dynamics of the earth, including field data from the First International Satellite Land Surface Climatology Project (ISLSCP) Field Experiment (FIFE) and the Oregon Transect Ecosystem Research (OTTER) Project. URL *<http://www-eosdis.ornl.gov/>*.

• **Socioeconomic Data and Applications Center (SEDAC).** Facilitates access to and integrates socioeconomic and earth science data, including administrative boundary data, census and household survey data, agricultural and land-use/land-cover data, international environmental treaties and associated status information, human health data, and associated metadata. (See also item VI.1.) URL *<http://sedac.ciesin.org/>*.

AVAILABILITY: For a current list of the DAACs and their access information, see URL *<http://eos.nasa.gov/daacsites.html>*. DAAC data holdings are described in the DAAC Science Data Plan database at *<http://spsosun.gsfc.nasa.gov/spso/sdp/sdphomepage.html>*.

IV.B-5 **National Cartography and Geospatial Center (NCG).** NCG is the U.S. Department of Agriculture's Natural Resources Conservation Service (NRCS) data archive and distribution center for spatial data and mapping products and services. AVAILABILITY: URL *<http://www.ftw.nrcs.usda.gov/ncg/ncg.html>*.

IV.B-6 **National Center for Atmospheric Research (NCAR).** NCAR is a research center operated by the University Corporation for Atmospheric Research (UCAR), which has extensive holdings of atmospheric and oceanographic data. The Web site provides access to real-time atmospheric and weather data from satellites, radar, and other sources (see Section V) and to various research data archives. AVAILABILITY: URL <*http://www.ncar.ucar.edu/*>.

IV.B-7 **National Remote Sensing Agency (NRSA) Data Centre.** This is the data center for the Indian Remote Sensing Satellite (IRS)-1C, which provides very-high-resolution panchromatic data. AVAILABILITY: URL <*http:// www.stph.net:80/nrsa/*>. Also see IRS-1C product information at URL <*http:// www.spaceimage.com/home/products/carterra/irs/irs_prod.html*>.

IV.B-8 **National Space Development Agency (NASDA) Earth Observation Center (EOC).** The EOC is a field center of the Japanese NASDA, established to develop satellite remote sensing technology. This Web site provides information on a range of satellites, including the Advanced Earth Observing Satellite (ADEOS/MIDORI), the Marine Observation Satellite (MOS)-1, the Japan Earth Resources Satellite (JERS)-1, the Tropical Rainfall Measuring Mission (TRMM), the Système pour l'Observation de la Terre (SPOT), Landsat, and the European Remote Sensing Satellites (ERS). It includes a range of satellite imagery concerning volcanoes, land-use change, weather, and other phenomena; 7 months of data from the ADEOS/MIDORI platform (November 1996-June 1997); and access to a test release of the Earth Observation Information System (EOIS) WWW Image Catalogue Service. (See also items II.B-4, III.B-3, and IV.B-9.) AVAILABILITY: URL <*http://www.eoc.nasda.go.jp/*>.

IV.B-9 **National Space Development Agency (NASDA) Earth Observation Research Center (EORC).** The EORC is a research center of the Japanese NASDA that provides access to data and information regarding various existing and planned Japanese satellites, including the Marine Observation Satellite (MOS)-1, the Japanese Earth Resources Satellite (JERS)-1, the Tropical Rainfall Measuring Mission (TRMM), and the Advanced Land Observing Satellite (ALOS). The Web site includes a data gallery and data library and links to both U.S. and Japanese Internet search engines. (See also items II.B-4, III.B-3, and IV.B-8.) AVAILABILITY: URL <*http://mentor.eorc.nasda.go.jp/index.html*>.

IV.B-10 **NOAA Satellite Active Archive.** The Satellite Active Archive is an online service of the U.S. National Oceanic and Atmospheric Administration's (NOAA) National Environmental Satellite, Data, and Information Service (NESDIS). It provides direct access to data from the Television InfraRed Observation Satellite (TIROS) Operational Vertical Sounder (TOVS), Defense Meteorological Satellite Program, and Advanced Very High Resolution Radiometer (AVHRR). (See also item III.A-8.) AVAILABILITY: URL <*http:// www.saa.noaa.gov/*>.

IV.B-11 **Space Monitoring Information Support (SMIS) Laboratory.** This laboratory of the Space Research Institute in Moscow provides access to data from the Geostationary Operational Meteorological Satellites (GOMS),

RESURS 01, and U.S. National Oceanic and Atmospheric Administration (NOAA) satellites. AVAILABILITY: URL *<http://smis.iki.rssi.ru/welcome.html>*.

V. OTHER NOTEWORTHY RESOURCES RELATED TO EARTH REMOTE SENSING APPLICATIONS AND EDUCATION

V-1 **Airborne Visible InfraRed Imaging Spectrometer (AVIRIS) Flight Locator.** AVIRIS is an optical sensor that delivers calibrated images of the upwelling spectral radiance in 224 contiguous bands with wavelengths from 400 to 2500 nanometers (nm). The instrument flies aboard a NASA ER-2 airplane at approximately 20 km above sea level. The Flight Locator illustrates the mean geographical path of AVIRIS flights (1992-1997) using the Topologically Integrated Geographic Encoding Reference (TIGER) Mapping Server developed by the U.S. Bureau of the Census. AVAILABILITY: URL *<http://makalu.jpl.nasa.gov/cgi-bin/tigermv.cgi>*.

V-2 **Center for Remote Sensing and Spatial Analysis (CRSSA), Rutgers University.** CRSSA's Web site provides information on a range of remote sensing and geographic information systems (GIS) projects, including the Global GRASS and GlobalARC CD-ROM Project, a cooperative activity with the U.S. Army Corps of Engineers Construction Engineering Research Laboratory (USA/CERL). The five Global GRASS CD-ROMs contain a wide range of global data sets related to land, oceans, atmosphere, biosphere, and human activities. GlobalARC contains 84 themes and a total of 147 raster layers (grids) with cell resolutions of 4'48" or finer. Other CRSSA projects of interest include work related to the use of remote sensing in archaeology and mountain gorilla protection. AVAILABILITY: URL *<http://deathstar.rutgers.edu/crssa.html>*.

V-3 **Core SW ImageNet®.** Core Software Technology is a commercial vendor of satellite data. It provides access to Landsat 4 and 5 data from Earth Observation Satellite Corporation (EOSAT); SovInformSputnik (SIS) data from the Russian Space Agency; Système pour l'Observation de la Terre (SPOT) 1, 2, and 3 data from SPOT IMAGE Corporation; polar orbiting data from the Australian Centre for Remote Sensing (ACRES); and European data from Eurimage. AVAILABILITY: URL *<http://www.coresw.com/>*.

V-4 **Declassified Intelligence Satellite Imagery.** A broad selection of satellite imagery from the 1960s and 1970s has been declassified by the U.S. intelligence community. Images may be browsed and ordered through the Earth Resources Observation System (EROS) Data Center (see item IV.B-4). AVAILABILITY: URL *<http://edcwww.cr.usgs.gov/dclass/dclass.html>*.

V-5 **Defense Meteorological Satellite Program (DMSP) Home Page.** This archive provides access to data from the DMSP satellites, including measurements from the Operational LineScan (OLS) instrument, the Special Sensor Microwave/Imager (SSM/I), and other instruments. AVAILABILITY: URL *<http://web.ngdc.noaa.gov/dmsp/>*.

V-6 **Delft Institute for Earth-Oriented Space Research.** Located at the Delft University of Technology in The Netherlands, this institute focuses on orbit determination, altimetry, and satellite geodesy. The Web site contains information on the European Remote Sensing Satellites ERS-1 and ERS-2, including the latest results from ERS-2 related to sea-surface variability, wind speed, wave height, and Gulf Stream velocities. AVAILABILITY: URL *<http://dutlru8.lr.tudelft.nl/>*.

*V-7 **EarthRISE: Earth Remote Imagery for Science and Education.** This database provides access to a collection of photographs of the earth taken from the NASA Space Shuttle during the past 15 years using hand-held cameras. The database interface supports searching by form, political unit, and topographical features. AVAILABILITY: URL *<http://earthrise.sdsc.edu/>*.

V-8 **Earth Satellite Corporation (EarthSat).** The EarthSat home page describes a range of EarthSat applications related to geology, flooding, agriculture, environment, transportation, and weather. AVAILABILITY: URL *<http://www.earthsat.com>*.

V-9 **EarthWatch, Inc.** EarthWatch provides high-resolution imagery, digital terrain models, and digital maps through its Digital Globe™ database. EarthWatch is planning to build and launch two commercial high-resolution imaging satellites, EarlyBird and QuickBird. EarlyBird was successfully launched on December 24, 1997. The Web site includes a short primer on remote sensing. AVAILABILITY: URL *<http://www.digitalglobe.com>*. Also see URL *<http://www.digitalglobe.com/rsensing/rsensing.html>*.

V-10 **Ecosystem Science and Technology Branch, NASA Ames Research Center.** The Ecosystem Science and Technology Branch hosts a number of research efforts involving applications of remote sensing, including the Center for Health Applications of Aerospace Related Technologies (CHAART), the Global Monitoring and Human Health Program, the Landsat Program, temporal urban mapping, use of synthetic aperture radar (SAR) to investigate high-latitude wetlands, and new sensor development. AVAILABILITY: URL *<http://geo.arc.nasa.gov/sge.html>*.

*V-11 **ENTRI Remote Sensing Gallery**. As part of the on-line guide for the Environmental Treaties and Resource Indicators (ENTRI) service, the Socioeconomic Data and Applications Center (SEDAC) has developed examples of the use of remote sensing to monitor compliance with international environmental treaties. AVAILABILITY: URL *<http://sedac.ciesin.org/entri>*.

V-12 **Eurimage.** Eurimage is a consortium of four European earth observation companies: British Aerospace (U.K.), Dornier (Germany), SSC-Satellitbild (Sweden), and Nuova Telespazio (Italy). It provides access to current data from a range of platforms, including Landsat, European Remote Sensing Satellites ERS-1 and -2, Television InfraRed Observation Satellite (TIROS), RESURS-01, Japan Earth Resources Satellite (JERS)-1, Kosmos (KVR 1000, TK-350, and MK-4), and the Russian MIR Space Station, as well as archived data from the

Coastal Zone Color Scanner (CZCS), the Heat Capacity Mapping Mission (HCMM), the Marine Observing Satellite (MOS), Seasat Synthetic Aperture Radar (SAR), and the Metric Camera. The Web site provides detailed technical information, a discussion of applications, and access to the EiNet on-line catalog (by subscription). AVAILABILITY: URL <http://www.eurimage.it/ >.

V-13 **The GeoTIFF Web Page.** This Web site provides information on the effort to establish an interchange format based on the Tag Image File Format (TIFF™) or georeferenced raster imagery. Last updated in June 1996, the site provides access to the official release version of the GeoTIFF Specification and lists GeoTIFF data providers and software vendors. Links are provided to a GeoTIFF File Transfer Protocol (FTP) archive maintained by the SPOT IMAGE Corporation. AVAILABILITY: URL <http://home.earthlink.net/~ritter/geotiff/ geotiff.html>.

*V-14 **The Great Globe Gallery.** This site provides more than 100 different links to images of the earth drawn from diverse sources. AVAILABILITY: URL <http://hum.amu.edu.pl/~zbzw/glob/glob1.htm>.

V-15 **Ionia "1 km AVHRR Global Land Data Set" Net-Browser.** This service of the European Space Agency (ESA) European Space Research Institute (ESRIN) allows users to browse Advanced Very High Resolution Radiometer (AVHRR) land data from 1992 to the present. The Web site also provides access to a Fire Atlas (fire detection based on AVHRR data) and to the Global 1 km AVHRR Server. AVAILABILITY: URL <http://shark1.esrin.esa.it/>. Also see URL <http://atlas.esrin.esa.it:8000/>.

V-16 **Naval Research Laboratory (NRL) Monterey Satellite Products.** The Naval Research Laboratory in Monterey, California, provides near-real-time access to a range of meteorological satellite data products, including data from the Geostationary Operational Environmental Satellites (GOES)-8 and -9 and the Geostationary Meteorological Satellite (GMS)-5. The site includes imagery, movies, documentation, forecasts, and other resources. AVAILABILITY: URL <http://www.nrlmry.navy.mil/projects/sat_products.html>.

V-17 **NOAA/NASA Pathfinder Programs.** The U.S. National Oceanic and Atmospheric Administration and NASA have sponsored a range of "pathfinder" activities to reprocess existing satellite data in support of global change research. Pathfinders include data based on the Advanced Very High Resolution Radiometer (AVHRR), the Television InfraRed Observation Satellite (TIROS) Operational Vertical Sounder (TOVS), the Geostationary Operational Environmental Satellites (GOES), the Special Sensor Microwave/Imager (SSM/I), and Landsat. AVAILABILITY: URL <http://xtreme.gsfc.nasa.gov/pathfinder/path_sites.html>.

V-18 **ORBIMAGE.** ORBIMAGE provides data from the OrbView 1 and 2 satellites and the planned OrbView 3 satellite. OrbView 1 includes two sensors, the Optical Transient Detector (OTD) and the Global Positioning System Meteorology (GPS/MET) instrument. OrbView 2 (SeaStar™) was successfully launched on August 1, 1997, carrying the Sea-Viewing Wide-Field-of-View Sensor (SeaWiFS) (see also item V-26). OrbView 2 data will be made available through an on-line OrbNet^SM Digital Archive after launch. The ORBIMAGE Web site also provides access to the SunCast® Data Service, which offers next-day forecasts of ultraviolet radiation to the general public. AVAILABILITY: URL <http://www.orbimage.com>.

V-19 **RADARSAT Program Official WWW Server.** The RADARSAT Program of the Canadian Space Agency maintains a WWW server that provides general information on the RADARSAT mission and links to sites that are distributing and using RADARSAT data. AVAILABILITY: URL <http://radarsat.space.gc.ca/>.

V-20 **Remote Sensing and Modeling.** The Center for Energy Studies at the Ecole des Mines de Paris has a Remote Sensing and Modeling group that focuses on earth observation from space and the modeling of processes contributing to the signal measured by spaceborne sensors. Applications include urban climate and air quality, urban mapping, oceanography and coastal zones, and meteorology and solar meteorology. AVAILABILITY: URL <http://www-cenerg.cma.fr/eng/tele/>.

V-21 **Satellite Imagery, Rosenstiel School of Marine and Atmospheric Science.** This site provides near-real-time High Resolution Picture Transmission (HRPT) and Local Area Coverage (LAC) daily image composites for selected marine areas in support of the Joint Global Ocean Flux Study (JGOFS). AVAILABILITY: URL <http://www.rsmas.miami.edu/images.html>.

V-22 **Satellite Images in the Network Office Archive.** This collection includes multiple satellite images of Long Term Ecological Research (LTER) sites, primarily from the Landsat Thematic Mapper (TM) and Système pour l'Observation de la Terre (SPOT) from the late 1980s and early 1990s. AVAILABILITY: URL <http://lternet.edu/lterdocs/catalog/tm.html>.

*V-23 **Satellite Observations of the Atmosphere.** The Mesoscale and Microscale Meteorology Division at the National Center for Atmospheric Research (see item IV.B-6) has developed a Web page on satellite observations of the atmosphere. The site includes access to real-time imagery from the Geostationary Operational Environmental Satellite (GOES)-8, an image gallery including selected animations, and a guide to satellite coverage and orbits (both geostationary and polar orbiting). AVAILABILITY: URL <http://www.mmm.ucar.edu/pm/satellite/satellite.html>.

V-24 **Satellite Remote Sensing and Archaeology.** This Web site, developed by M. J. F. Fowler in the United Kingdom, provides samples of satellite imagery used in remote sensing, a calendar of relevant meetings, a bibliography, and links to other related resources. AVAILABILITY: URL <http://ourworld. compuserve.com/homepages/mjff/homepage.htm>.

*V-25 **SatPasses: Predictions of Satellite Passes over North American Cities.** This Web site delivers predictions of when the Space Shuttle, the MIR Space Station, and other satellites will pass over selected North American cities. It includes links to other satellite tracking resources (including a comparable service for cities in other regions of the world). AVAILABILITY: URL *<http://www.bester.com/satpasses.html>*.

*V-26 **SeaWiFS Project Home Page.** This home page contains a wide range of data and information resources about the Sea-viewing Wide-Field-of view Sensor (SeaWiFS) mission (launched in August 1997) (See item V-18.) Included are technical documents, browse imagery, a teacher's guide, lists of ground stations, and current mission information. AVAILABILITY: URL *<http://seawifs.gsfc.nasa.gov/SEAWIFS.html>*.

V-27 **Space Imaging EOSAT's Online CARTERRA™ Archive.** This commercial digital archive contains panchromatic, multispectral, and wide-field data from the Indian Remote Sensing Satellite IRS-1C, high-resolution digital photography, and selected radar and submeter products. AVAILABILITY: URL *<http://www.spaceimage.com/home/browse/index.html>*.

V-28 **Space Shuttle Earth Observations Project (SSEOP) Database of Photographic Information and Images.** This is an on-line database of photographs taken from the Space Shuttle using hand-held cameras. The database may be searched by feature and location and returns a jpeg image and text description of Shuttle photographs. Information on ordering hard-copy photographs is also provided. AVAILABILITY: URL *<http://eol.jsc.nasa.gov/sseop/>*.

V-29 **Space Science and Engineering Center (SSEC) Real-Time Data.** The SSEC at the University of Wisconsin at Madison provides real-time data from geostationary satellites, such as the Geostationary Operational Environmental Satellites (GOES). This site contains browse-quality images only. AVAILABILITY: URL *<http://www.ssec.wisc.edu/data/>*.

V-30 **SPOT IMAGE.** The French Système pour l'Observation de la Terre (SPOT) satellites 1, 2, and 3 have provided high-resolution, stereo imagery since 1986. SPOT 4 is scheduled for launch in early 1998. SPOT IMAGE Corporation offers a range of data products, including land classification data and high-resolution urban digital imagery (both satellite and aerial). The Worldwide SPOT Scene Catalogue is available on CD-ROM, and an online catalog, DALI, provides metadata for more than 4 million SPOT scenes and browse images for more than 1 million. AVAILABILITY: URL *<http://www.spot.com>*.

V-31 **Stratospheric Ozone and Human Health Project Home Page.** This on-line service provides access to historical and near-real-time estimates of surface doses of ultraviolet radiation based on satellite measurements from the Total Ozone Mapping Spectrometer and other instruments. The site includes an online bibliography on stratospheric ozone depletion, ultraviolet radiation, and human health and provides links to other related data and Internet resources. AVAILABILITY: URL *<http://sedac.ciesin.org/ozone/>*.

*V-32 **View from Satellite.** This utility provides a view of the earth from the current position of the selected satellite in its orbit. Orbital data for hundreds of earth observing and communications satellites are updated regularly. AVAILABILITY: URL *<http://www.fourmilab.ch/earthview/satellite.html>*.

VI. SELECTED E-MAIL/WWW GATEWAYS

VI.1 E-Mail Access to WWW Human Dimensions Data and Information Resources. The Consortium for International Earth Science Information Network (CIESIN) Socioeconomic Data and Applications Center (SEDAC) maintains a service to provide e-mail access to human dimensions data and information resources available over the WWW (see also item IV.B-4). Users may retrieve WWW documents either as plain text or in hypertext format for local viewing with a standard browser. The service currently utilizes the Agora software developed by the W3 Consortium, which does not yet support hypertext forms. The service is free, but registration is required. AVAILABILITY: Send an e-mail message to *<www.mail@ ciesin.org>* with the word "help" in the body of the message. For assistance, contact CIESIN User Services at *<ciesin.info @ciesin.org>*.

VI.2 GetWeb. SatelLife HealthNet's GetWeb server allows users to request WWW pages via e-mail. It supports both text and hypertext format documents and is capable of submitting keyword searches to standard Internet search services. SatelLife is an international not-for-profit organization serving health communication and information needs of developing countries. AVAILABILITY: URL *<http://www.healthnet.org/dist/getweb/>*, or send e-mail to *<getweb@usa.healthnet. org>* with the subject line blank and the command "GET <http://www.healthnet.org/ dist/getweb/help/gwhelp.HTML" in the body of the message.

ACKNOWLEDGMENTS

Selection and descriptions of resources represent the views of the author and do not necessarily represent the views of the Consortium for International Earth Science Information Network (CIESIN), the Socioeconomic Data and Applications Center (SEDAC), or the U.S. National Aeronautics and Space Administration (NASA). This work was supported by NASA Contract NAS5-32632. The author thanks Marie-Lise Shams for assistance in verifying and updating the guide and preparing the online version.

Appendix B

Glossary

Mark Patterson

NOTE: italicized terms in the definitions are defined in this glossary.

Albedo. The ratio of the amount of *electromagnetic radiation* reflected by a surface to the amount falling on a surface. Lighter-colored surfaces, such as snow, have higher albedo ratios than do darker-colored surfaces, such as trees.

Amazonia. A term denoting the legal boundary of several Brazilian states that encompass a portion of the Amazon River Basin in Brazil.

ARC/INFO. *Geographic information system* software produced by the Environmental Research Systems Institute (ERSI).

AVHRR: Advanced Very High Resolution Radiometer. A U.S. National Oceanic and Atmospheric Administration satellite that collects *multispectral* data. This satellite has a *spatial resolution* of 1.1 × 1.1 km and a *temporal resolution* of approximately 12 hours.

Cadastral map. A map of the extent, value, and ownership of land, originally used as a basis for taxation.

Census tract. A small areal unit used in collecting and reporting census data.

Central business district. The location in an urban area where the concentration of commercial activity is most dense.

Centrifugal forces. A term used to describe factors, such as crime and pollution, that push people and businesses away from the center of urban areas.

Centripetal forces. A term used to describe factors, such as employment opportunities, that pull or attract people and businesses toward the center of urban areas.

Commission error. A measure of classification accuracy, based on the number of *pixels* incorrectly assigned to a particular class that belong in other classes. It is the probability that a pixel in an image is not actually representative of a class in reality.

Composite image. An image derived from multiple images of the same area. A false color composite is a common composite image in which different spectral bands of the same image are displayed simultaneously using different colors.

DEM: Digital elevation model. A digital file in *raster* format in which each georeferenced *pixel* has an elevation value.

Digital number. The numerical value assigned to a *pixel* in a digital image.

Doppler radar. A radar system that measures the velocity of an object by recording the change in frequency of returning radar waves caused by the object's moving.

ENSO: El Niño/Southern Oscillation. An atmospheric and oceanic phenomenon that affects ocean currents. Cool ocean currents are replaced by warm ocean currents in the eastern tropical Pacific Ocean. The change in the temperature of ocean currents affects precipitation patterns on land, which in turn alters vegetation growth and agriculture. ENSO events occur roughly every 4 to 6 years. The Southern Oscillation refers to fluctuations in atmospheric pressure and rainfall across the Indo-Pacific region.

Electromagnetic spectrum. A continuous sequence of electromagnetic radiation arranged according to wavelength. The spectrum also includes visible light.

Epistemology. The study of knowledge. More specifically, it is the study of the nature of and grounds for knowledge, especially with reference to the limits and validity of knowledge.

FEWS: Famine Early Warning System. A U.S. Agency for International Development progam that monitors a variety of social and physical variables in a country or region of Africa in order to identify populations at risk of food insecurity. Advance warning can then be provided to the U.S. government so it will be prepared to commit food-aid resources before a crisis occurs. *Remote sensing* is an integral part of FEWS, as it provides a composite indicator of vegetation growth and rainfall distribution.

Fractal. Short for fractional dimension. A geometrical or physical structure that has an irregular or fragmented shape at all scales of measurement. Coastlines are often depicted or represented by fractals.

Fuzzy boundary. A boundary that is treated as a band of uncertainty.

GCM: general circulation model. A term used to describe a computer model that simulates large-scale features of atmospheric circulation by solving a set of equations governing atmospheric motion. Changes in temperature and precipitation are typical outputs of such a model.

GIS: geographic information system. A computer system consisting of hardware and software used to store, manipulate, analyze, and display *georeferenced* data.

Geometric correction. An image preprocessing procedure that corrects *spatial distortions*. Features in an image are repositioned so their locations are correct based on a known coordinate system, such as latitude/longitude.

Georeferencing. The process of referencing elements in an image to a known coordinate system, such as latitude/longitude or Universal Transverse Mercator (UTM).

GOES: Geostationary Operational Environmental Satellites. U.S. National Oceanic and Atmospheric Administration satellites that travel at the same speed and direction as the earth's rotation. They are used to collect weather data.

Global carbon cycle. The global circulation of carbon. Carbon can be transferred from the ground (by burning trees and fossil fuels, for example) to the atmosphere (in the form of the greenhouse gas carbon dioxide) and back to the earth (by photosynthesis). Both human actions and natural phenomena affect the amount of carbon in the atmosphere, mainly in the form of the carbon dioxide.

GPS: Global Positioning System. A system of 24 satellites that orbit the earth and are used in computing a location on earth. A GPS receiver unit tracks these satellites and using geometry can compute the GPS receiver's location and elevation. GPS units are commonly used in collecting *ground-referenced data.*

Ground-referenced data. *Georeferenced* data that are collected at the actual ground location of an area encompassed by imagery. A *GPS* unit is typically used to provide the georeferencing.

Ground truthing. Also called ground referencing. In remote sensing, it is the exercise of field work to verify the interpretation of imagery.

Human ecology. The study of humans' interaction with their physical and social environment.

IRS: Indian Remote Sensing. A near-polar, sun-synchronous satellite system, developed by National Natural Resource Management Systems (India), that collects *multispectral* data of the earth with a *spatial resolution* of 72.5 × 72.5 m.

Landsat. An unmanned satellite system initially operated by the U.S. National

Aeronautics and Space Administration and now by the private firm, Earth Observation Satellite Company. These satellites collect *multispectral* data. Presently, only Landsat 5 is still functioning. Landsat 6 failed to achieve orbit, and Landsat 7 is expected to be launched in 1998.

Landsat MSS: Landsat Multispectral Scanner. An imaging system found on the first five *Landsat* satellites. The system collects *multispectral* data in four *nonthermal radiation* bands with a spatial resolution of 79 × 79 m. On Landsat 1 to 3, data were also collected in a single thermal band with a *spatial resolution* of 250 × 250 m.

Landsat TM: Landsat Thematic Mapper. A multispectral scanner imaging system on board the *Landsat* 4 and 5 satellites. The imaging system collects *multispectral* data in seven bands (six nonthermal and one thermal) ranging from *visible radiation* to *thermal infrared radiation*. The nonthermal bands have a *spatial resolution* of 30 × 30 m, whereas the thermal band has a spatial resolution of 120 × 120 m. The *temporal resolution* is 16 days.

Latent heat flux. The increase in internal energy of a substance associated with a change in molecular configuration (e.g., liquid to gas) over a given period of time. Latent heat flux is critical for maintaining global energy balance as it transfers heat to and from the earth's surface.

Markov chain analysis. A statistical analysis technique in which the probability of an event in a sequence of random events is affected by the outcome of the most recent event.

Maximum likelihood classifier. A *supervised classification* technique based on the probability density function (distribution) of the spectral signatures of *pixels*, used to train the computer to assign pixels to classes. The probability of a pixel's belonging to each distribution is computed, and the pixel is assigned to the distribution (class) with the highest probability.

Mesophyll reflectance. The *reflectance* of *near-infrared radiation* from the spongy mesophyll (internal cell structure) found in leaves.

Metadata. The documentation associated with a data set, including a description of the data, the date the data were collected, and a source.

Mid-infrared radiation. The portion of the *electromagnetic spectrum* spanning 1.55 to 3 micrometers (μm). Mid-infrared radiation is sensitive to the moisture content of features in an image.

Monte Carlo simulation. A term used to describe the simulation or modeling of an event whose occurrence is random and not predicated on previous events. It alludes to games of chance, such as craps or roulette, played in Monte Carlo, in which the outcome of events is not dependent on previous outcomes. Monte Carlo simulation is often used to generate random numbers for use in computer modeling.

Multispectral scanner. An imaging system, such as the Landsat Multispectral

Scanner (MSS), that simultaneously collects data from the same scene at different wavelengths in the *electromagnetic spectrum.*

NDVI: Normalized Difference Vegetation Index. A measure of vegetation vigor computed from *multispectral* data. It is computed by subtracting the red band from the *near-infrared* band and dividing the result by the sum of the red and near-infrared bands.

Nearest neighbor resampling. The reassignment of a *digital number* to a pixel by assigning the value of the pixel nearest to the location of the resampled pixel. In geometric corrections of images, values must be assigned to the pixels in the new coordinate system. This process is known as resampling.

Near-infrared radiation. The portion of the *electromagnetic spectrum* spanning 0.7 to 1.55 micrometers (mm). This radiation is reflected by vegetation and hence is a good indicator of vegetation content in an image.

Nonthermal radiation. Radiation that is reflected by an object and is found in the visible to the mid-infrared (0.45-3.00 mm) portion of the *electromagnetic spectrum.*

Omission error. A measure of classification accuracy based on the number of *pixels* incorrectly excluded from a particular class. It is the probability of a ground reference pixel's being classified incorrectly.

Orthophotograph. A digital aerial photograph that has been *geometrically corrected* for distortions.

Panchromatic (pan) scanner. An imaging system that records all visible light in a single band. This type of scanner is found on board the *SPOT* satellites.

Phenology. The study of the timing of recurring natural phenomena in the life cycle of plants. The different stages of the annual cycle of a plant have implications for vegetation classification from remotely sensed data and change detection analysis.

Photogrammetry. The science and technology of procuring reliable measurements using aerial photographs.

Pixel. The smallest unit of spatial resolution in a photo or remotely sensed image. It refers to the area on the ground (*spatial resolution*) represented by a digital number. A remotely sensed digital image may be composed of millions of pixels. Pixel size varies in accordance with the type of sensor used.

Polygon. A closed-plane feature formed by a bounded series of straight lines. A polygon may be any shape.

RADARSAT. A Canadian Space Agency satellite that collects C-band active microwave imagery. C-band microwaves have longer wavelengths than visible and infrared waves and range from 3.8 to 7.5 centimeters. They are useful for penetrating features such as clouds, tree canopies, and sand to record what lies underneath.

Radiometric correction. An image processing procedure in which corrections are made to inaccurate *digital numbers* resulting from sensor degradation or malfunctioning.

Radiometry. The science and techniques involved in using radiometers. A radiometer is a device capable of detecting *electromagnetic radiation*.

Raster. A spatial data model in which features are represented by *pixels*. Each pixel is assigned a value that corresponds to a feature. Data from remotely sensed images are stored in raster format.

Reflectance. The ratio of energy reflected by a surface to the total amount of energy striking the surface.

Remote sensing. The use of *electromagnetic radiation* sensors to record images of an environment. Black-and-white photographs and satellite imagery are types of remotely sensed data.

Rent theory. A theory according to which land use will be such that it returns the highest possible amount of profit.

Semivariogram. A geostatistical graph that displays similarities (or differences) between two or more spatially discrete phenomena by measuring the variance between the phenomena. Variance is assumed to be represented by the distance between the phenomena, so that phenomena closer to one another have a smaller variance.

Senescence cycle. The phases of plant growth from maturity to death that are characterized by an accumulation of metabolic products (leaves, seeds, and fruit), increase in respiratory rate, and loss in dry weight (especially in the leaves). The spectral signature of a plant will vary depending on the phase of the cycle.

Sensible heat flux. A measure of the kinetic energy of the molecular motion of a substance (e.g., air) over a given time period. A thermometer is used to measure sensible heat (i.e., air temperature).

Sensitivity analysis. A type of analysis that accounts for the dependence of an outcome variable on a causal factor by providing a range of possible outcomes based on changes in the value of the causal factor. For example, a range of precipitation values may be computed in a *general circulation model* by changing the amount of carbon dioxide (causal factor) in the atmosphere each time the model is executed.

Solar insolation. Energy in the form of electromagnetic waves from the sun that are intercepted by the earth.

Spatial dependency. A condition arising when a strong relationship exists between an event at one place and events in other places.

Spatial distortion. Changes in the position and shape of features in an image with respect to their true position and shape.

Spatial resolution. The level of detail that can be detected by a sensor, often measured simply by the dimensions of a *pixel* in an image. The finer the spatial resolution, the more information can be discerned from an image. *Landsat TM* has a spatial resolution of 30 × 30 m, while *SPOT PAN* has a spatial resolution of 10 × 10 m.

SPOT: Système pour l'Observation de la Terre. A French commercial satellite program designed to collect high-*spatial-resolution* panchromatic (10 × 10 m) and *multispectral* (20 × 20 m) images.

SPOT HRV: SPOT High Resolution Visible. The scanner system used on board the *SPOT* satellites.

SPOT MX (usually referred to as SPOT XS). SPOT *multispectral scanner* imaging system that collects data in the green, red, and near-infrared portions of the *electromagnetic spectrum*. SPOT XS images have a *spatial resolution* of 20 × 20 m.

SPOT PAN: SPOT Panchromatic. A SPOT imaging system that collects data in a single spectral band. SPOT PAN images have a *spatial resolution* of 10 × 10 m.

Stereoscopic imagery. Imagery or aerial photographs composed of two partially overlapping images that appear three-dimensional when viewed through stereoscopic glasses.

Supervised classification. A classification technique in which the operator provides classification information used by the computer to assign *pixels* to classes.

Swidden cultivation. Also known as swidden agriculture or more colloquially as slash-and-burn agriculture. It is a form of agriculture that involves clearing forests or bushland for agricultural purposes. When soil fertility has decreased after a few years, the land goes to fallow of variable length before being cleared again.

Temporal resolution. The time it takes for a satellite to return to or revisit the same location. *Landsat TM*, for example, has a temporal resolution of 16 days.

TIR: thermal infrared radiation. The portion of the *electromagnetic spectrum* spanning 3 to 14 micrometers (μm). Objects that emit heat can be detected in this portion of the spectrum.

TIGER file: Topologically Integrated Geographic Encoding Reference file. A digital geographic coding system developed by the U.S. Bureau of the Census.

TIN: triangulated irregular network. A network of triangular facets drawn among points on a surface. The coordinates and elevations of the points forming the vertices of each triangular facet are used to compute the slope and aspect (direction) of the terrain.

Topographic distortions. Distortions in reflectance readings by a satellite caused by topographic variability. Shadows are the most common form of topographic distortions.

Unsupervised classification. A computer algorithm that assigns *pixels* to classes with no prior instructions from the operator. Classes are defined such that differences among them are maximized, while differences within them are minimized.

USGS Level I class. A U.S. Geological Survey hierarchical land-use classification scheme based on a level of spatial resolution that can be achieved by satellite imagery, such as *Landsat* (scale less than 1:250,000). An example of a USGS Level I class is urban land use.

USGS Level II class. A U.S. Geological Survey hierarchical land-use classification scheme based on a level of spatial resolution that can be achieved by high-altitude imagery (scale 1:80,000 to 1:250,000). An example of a USGS Level II class is residential land use.

USGS Level III class. A U.S. Geological Survey hierarchical land-use classification scheme based on a level of spatial resolution that can be achieved by medium-altitude imagery (scale 1:20,000 to 1:80,000). An example of a USGS Level III class is single-family-dwelling land use.

USGS Level IV class. A U.S. Geological Survey hierarchical land-use classification scheme based on a level of spatial resolution that can be achieved by low-altitude imagery (scale greater than 1:20,000).

Vector. A spatial data model in which features are represented by points, lines, and *polygons*. A point could represent a city, a line could represent a road, and a polygon could represent a forest.

Visible radiation. The portion of the *electromagnetic spectrum* spanning 0.4 to 0.7 micrometers (μm). Radiation in this portion of the spectrum is visible to the human eye in the form of different colors.

Wetness Vegetation Index. A feature produced by the Kauth-Thomas linear transformation (also called the tasseled cap transformation) that displays vegetation and soil moisture content.

Biographical Sketches of
Contributors and Editors

EDUARDO S. BRONDIZIO is a postdoctoral fellow and assistant director of the Anthropological Center for Training and Research on Global Environmental Change at Indiana University. His work has focused on agricultural and agroforestry intensification, land-use and land-cover change, secondary succession in the Eastern Amazonia region (especially in the Amazon estuary), and the application of remote sensing to these issues. Previously, he was the coordinator of the project that led to the 1990 publication *Atlas of the Atlantic Forest Remains in Brazil*. His work has appeared in numerous journals, including *Human Ecology, Photogrammetric Engineering and Remote Sensing*, and *Research in Economic Anthropology*. He received a PhD degree in environmental science from Indiana University.

APHICHAT CHAMRATRITHIRONG is the former director and an associate professor of the Institute for Population and Social Research, Mahidol University, Thailand. His current research has focused on population and environment, migration, and reproductive health in Thailand, including the evaluation of a condom promotion program and the Thailand National Contraceptive Prevalence Survey. He was a member of two research review committees of the World Health Organization. He is past president of the Thailand Population Association. He received a BA degree in political science from Chulalongkorn University, Thailand, and AM and PhD degrees in sociology from Brown University.

ROBERT CHEN is director of the Data Center Services Division at the Consortium for International Earth Science Information Network (CIESIN) and manager of the Socioeconomic Data and Applications Center (SEDAC), operated by CIESIN for the U.S. National Aeronautics and Space Administration (NASA). Previously, he served on the faculty at Brown University with the Alan Shawn Feinstein World Hunger Program, where his research dealt with global environmental change and future food security, hunger among refugees, and the measurement and alleviation of hunger. His current work focuses on the intersection between environment and security, the interdisciplinary integration of natural and social science data, and the human dimensions of global environmental change. He holds a bachelor's degree in earth and planetary sciences, master's degrees in meteorology and physical oceanography and in technology and policy from the Massachusetts Institute of Technology, and a PhD degree in geography from the University of North Carolina, Chapel Hill.

RINKU ROY CHOWDHURY is a doctoral student in the Graduate School of Geography at Clark University. Her specialties are geographic information systems and remote sensing, ecology, and human-environment relationships. She holds a BA degree in environmental science and computer science from Wellesley College and an MS degree in conservation ecology and sustainable development from the University of Georgia.

DAVID J. COWEN is director of the Liberal Arts Computing Lab, codirector of the NASA Visiting Investigator Program, and Carolina distinguished professor of geography at the University of South Carolina. His research and instruction have focused on geographic information systems. He is currently the president of the Cartographic and Geographic Information Society of the American Congress on Surveying and Mapping, and he has served on the Mapping Science Committee of the National Research Council and the Geographic Information Systems Commission of the International Geographical Union. He has also been chair of the Association of American Geographers' Geographic Information Systems Specialty Group and the South Carolina State Mapping Advisory Committee. He is author or coauthor of numerous publications dealing with advances in spatial data handling. He earned BA and MA degrees from the State University of New York at Buffalo and a PhD degree in geography from Ohio State University.

BARBARA ENTWISLE is professor of sociology and a fellow of the Carolina Population Center at the University of North Carolina at Chapel Hill. Her current research focuses on the population dynamics and demographic responses to changes in the landscape and environment and on spatial organization of the landscape in Nang Rong, Thailand, due to social and environmental forces. Her work has been published in numerous journals, including *Demography, Population Research and Policy Review, Studies in Family Planning*, and the *American Sociological Review*. She is currently a member of ther National Institutes of Health Social Science and Population Study Section, a member of the advisory board of the General Social Survey and about to become co-editor of *Demography*. She received an AB degree in sociology-anthropology from Swarthmore College and AM and PhD degrees in sociology from Brown University.

PAUL R. EPSTEIN is Associate Director of the Center for Health and the Global Environment at Harvard Medical School. He has worked in medical, teaching, and research capacities in Africa, Asia, and Latin America. He has coordinated and coedited an eight-part series on Health and Climate Change for *The Lancet* and is a principal core author for *Human Health and Climate Change*, a publication supported by the World Health Organization, the World Meteorological Organization, the U.N. Environmental Program, and the Intergovernmental Panel on Climate Change. He is currently coordinating an integrated assessment of disease events along the East Coast of North America, the Gulf of Mexico, and the Caribbean. He is a member of the Health of the Oceans Module of the Global Ocean Observing System. He received a BA degree from Cornell University, an MD degree from the Albert Einstein College of Medicine, and an MPH degree in tropical public health from the Harvard School of Public Health.

JACQUELINE GEOGHEGAN is an assistant professor in the Department of Economics and a particpating faculty member in the Environmental School at Clark University. Previously, she was a postdoctoral research fellow in the Department of Agricultural and Resource Economics at the University of Maryland. Her general fields of research are environmental economics, development economics, and environmental law and policy. Her specific research interests include the design and use of environmental taxes for regulatory and fiscal reform; spatially explicit theoretical and empirical modeling of human-induced land-use change (in both the United States and Mexico); and linkages among human behavior, urban transportation, and environmental quality. She received a PhD degree in agricultural and resource economics from the University of California, Berkeley.

CHARLES HUTCHINSON is a professor of arid lands studies at the University of Arizona. He serves as the director of the Arizona Remote Sensing Center, Office of Arid Lands Studies, College of Agriculture, and as chair of the campus-wide Interdisciplinary Committee on Remote Sensing and Spatial Analysis. His research has focused most recently on developing methods for famine early warning in Africa and the Caribbean Basin and working with the Global Information and Early Warning System of the Food and Agriculture Organization of the United Nations and the Famine Early Warning System and the Office of Foreign Disaster Assistance of the U.S. Agency for International Development. Previously, he held a Gilbert F. White fellowship at Resources for the Future and was a member of the technical staff at the Jet Propulsion Laboratory at the California Institute of Technology and at the EROS Program of the U.S. Geological Survey.

He received BA, MA, and PhD degrees in geography from the University of California, Riverside.

JOHN R. JENSEN is a Carolina distinguished professor in the Department of Geography at the University of South Carolina. His research has focused on remote sensing of vegetation biophysical resources, especially inland and coastal wetlands; urban/suburban land use and land cover; and the development of improved digital image processing classification, change detection, and error evaluation algorithms. He is a contributing author to the *Manual of Remote Sensing* and the *Manual of Photographic Interpretation.* He was the President of the American Society for Photogrammetry and Remote Sensing from 1995-1996. He has received several awards in the field of remote sensing, including the Alan Gordon Memorial Award for significant achievements in remote sensing and photographic interpretation from the American Society of Photogrammetry and Remote Sensing, and the Remote Sensing Medal from the Remote Sensing Specialty Group of the Association of American Geographers. Dr. Jensen received a BA degree in physical geography from the California State University at Fullerton, a master's degree from Brigham Young University, and a PhD degree from the University of California, Los Angeles.

DIANA LIVERMAN is director of the Latin American studies program at the University of Arizona, where she is also associated with the Department of Geography, the Institute for the Study of Planet Earth, the Office of Arid Lands Studies, and the Udall Center for Studies in Public Policy. She has published widely on drought, climate impacts, resource management, and environmental policy. Her current research examines the social causes and consequences of global and regional environmental change, especially the impacts of climate change and variability on water resources and agriculture in the Americas and the social causes of land-use and land-cover change in Mexico. She is cochair of the Scientific Advisory Committee of the Inter-American Institute for Global Change. She received a BA degree from the University of London, England, an MA degree from the University of Toronto, Canada, and a PhD degree from the University of California, Los Angeles, all in geography.

STANLEY A. MORAIN is professor of geography and director of the Earth Data Analysis Center at the University of New Mexico and has served as chair of the Geography Department. He has focused his basic and applied research on understanding the spectral and spatial attributes of natural resource development and management. He is the author or editor of several books and many technical papers on the subject. He is a fellow of both the American Association for the Advancement of Science and the American Society for Photogrammetry and Remote Sensing (ASPRS) and is a past national president of ASPRS. He currently serves as editor-in-chief of the ASPRS journal, *Photogrammetric Engi-*

neering and Remote Sensing, and is on the editorial boards of the *International Journal of Remote Sensing* and *GeoCarto International: A Multidisciplinary Journal of Remote Sensing*. Dr. Morain received a BA degree from the University of California, Riverside, and a PhD degree from the University of Kansas, both in geography.

EMILIO F. MORAN is professor of anthropology, director of the Anthropological Center for Training and Research on Global Environmental Change, codirector of the Center for the Study of Institutions, Population, and Environmental Change, and professor of environmental science at the School of Public and Environmental Affairs, all at Indiana University. He has carried out research in Amazonia for the past 25 years on migration, land use, social organization of frontier communities, and tropical ecology. In recent years he has begun applications of remote sensing to issues of land-use and land-cover change. He is currently studying the role that the structure of households may play in differential rates of deforestation. He is the author of numerous books, edited volumes, and journal articles. He received a PhD degree in anthropology from the University of Florida.

YELENA OGNEVA-HIMMELBERGER is a doctoral candiate in the School of Geography at Clark University. Previously, she was a researcher in the Institute of Geography of the Russian Academy of Sciences. Her specialty is geographic information systems, remote sensing, and global environmental change. She holds BS and MS degrees in physical geography from Moscow State University.

MARK PATTERSON is a student in the doctoral program in geography at the University of Arizona. His current research examines the social implications of using geographic information systems in resource management decision making. His other research interests include using remote sensing to classify forest fire severity for the U.S. Forest Service, examining land-cover change in Mexico through satellite image interpretation, and investigating the potential impacts of climate change on the resource base of the Southwest. Mr. Patterson received a BS degree in geography from University of Victoria, Canada, and an MS degree in geography from the University of Guelph, Canada.

LOWELL PRITCHARD, JR. is a research associate in the Tropical Conservation and Development Program and a doctoral student in the Food and Resource Economics Department, both at the University of Florida. He is also the science officer for Focus 1 of the International Geosphere-Biosphere Programme/International Human Dimensions Programme joint project on Land-Use and Land-Cover Change (LUCC). He participates in the Resilience Network of the Beijer International Institute for Ecological Economics, through his work on institutional change, network externalities, and human agency. He holds degrees in environmental engineering and systems ecology from the University of Florida and in zoology from Duke University.

RONALD R. RINDFUSS is professor of sociology and fellow of the Carolina Population Center at the University of North Carolina, Chapel Hill. His social demographic research has focused on fertility, household formation, marriage and marital dissolution, and child care, in both the Untied States and a number of Asian countries. More recently, he has been examining the link between population change and environmental change, with particular emphasis on the role of migration. With colleagues at the Institute for Population and Social Research, Mahidol University, and various American universities, he has been part of a longitudinal study of Nang Rong District, Thailand. He is a past president of the Population Association of America and a former director of the Carolina Population Center. He is a member of the International Scientific Planning Committee for the 1999 Open Meeting of the Human Dimensions of Global Environmental Change Research Community. He holds a PhD degree from Princeton University.

STEVEN SANDERSON is Vice President for Arts and Sciences and Dean of Emory College at Emory University. He has studied the politics of rural poverty, natural resource use, and environmental change, with special reference to Latin America. Previously, Dr. Sanderson served as Ford Foundation program officer for rural poverty and resources in Brazil, where he designed and implemented the foundation's Amazon program. He was also on the faculty of the University of Florida, where he directed the Tropical Conservation and Development Program and founded the Conservation and Development Forum, a worldwide partnership on sustainable futures funded by the Ford Foundation. Since 1994 he has chaired the Social Science Research Council Committee for Research on Global Environmental Change, and he serves as a member of the International Scientific Steering Committee of the International Geosphere-Biosphere Programme/International Human Dimensions Programme on Global Environmental Change project on Land-Use/Cover Change. He is an elected member of the board of directors of Oxfam America. He holds a PhD in political science from Stanford University.

THOMAS L. SEVER is a remote sensing archeologist at the Global Hydrology and Climate Center of the NASA-Marshall Space Flight Center in Alabama. Previously, he worked in the Earth Systems Science Office of the NASA-Stennis Space Center in Mississippi, on the faculty of the University of Southern Mississippi, and as a scientific supervisor at Lockheed Electronics Corporation. His work has focused on bringing remote sensing and geographic information systems technology to the disciplines of anthropology and archeology. He has been conducting research in northern Guatemala for a decade, and his work contributed to the establishment of the Maya Biosphere Reserve. He has also worked with airborne and satellite systems conducting international research in Israel, Peru, Chile, Mexico, Costa Rica, Guatemala, and the United States. His awards include the Society of Professional Archeologists Exceptional Achievement Award, the NASA Exceptional Achievement Award, and the NASA Exceptional Scientific Achievement Award. He holds a PhD in anthropology from the University of Colorado.

DAVID L. SKOLE is professor of geography and director of the Basic Science and Remote Sensing Initiative at Michigan State University. His research emphasis has been on developing numerical models of the global carbon cycle, focusing on the role of land cover and ecosystem dynamics. He has worked extensively with large satellite data sets and data collection systems, including Landsat and the Advanced Very High Resolution Radiometer. He serves on a number of national and international advisory panels, including the NASA/Earth Observing System-Data and Information System advisory panel, the LP-Distributed Active Archive Center Advisory Panel and the Oak Ridge Distributed Active Archive Center Advisory Panel. In addition, he is chair of the International Geosphere-Biosphere Programme/International Human Dimensions Programme on Global Environmental Change project on Land-Use/Land-Cover Change and a member of the Steering Committee for the International Geosphere-Biosphere Programme's Data and Information System, as well as the Science Advisory Panel for the Programme's Southeast Asian Regional Committee for the Global Change System for Analysis, Research, and Training (START).

PAUL C. STERN is study director of the Committee on the Human Dimensions of Global Change and the Committee on International Conflict Resolution at the National Research Council, research professor of sociology at George Mason University, and president of the Social and Environmental Research Institute. In his major research area, the human dimensions of environmental problems, he has written numerous scholarly articles and coedited and coauthored several books; he has also authored a textbook on social science research methods and coedited several books on international conflict issues. He is a fellow of the American Psychological Association and the American Association for the Advancement of Science. He holds a BA degree from Amherst College and MA and PhD degrees in psychology from Clark University.

B. L. TURNER II is the Milton P. and Alice C. Higgins professor of environment and society and director of the Graduate School of Geography at Clark University and former director of the George Perkins Marsh Institute. His research focus is nature-society relationships, subjects on which he has authored or edited several books and numerous articles, ranging from ancient Maya agriculture and environment in Mexico and Central America, to contemporary agricultural change in the tropics, to global land-use change. He is a member of the National Academy of Sciences and has been awarded honors for research by the Association of American Geographers and the Centenary Medal by the Royal Scottish Geographical Society. He received BA and MA degrees in geography from the University of Texas at Austin and a PhD degree in geography from the University of Wisconsin-Madison.

STEPHEN J. WALSH is professor of geography at the University of North Carolina, director of the spatial analysis laboratory in that department, a senior research fellow at the university's Sheps Center for Health Services Research, and director of the Spatial Analysis Unit at the Carolina Population Center. Previously, he was on the faculty of Oklahoma State University, where he established the Center for Application of Remote Sensing. His research focuses on population-environment interactions, land-use and land-cover change, environmental modeling, health care accessibility, and spatial analysis; his methodological research focuses on geographic information systems and remote sensing. Most recently he has worked on population-environment projects in Thailand and Ecuador and environmental modeling in Montana, North Carolina, and Georgia. He earned a PhD degree in geography from Oregon State University.

CHARLES H. WOOD is director of the Center for Latin American Studies at the University of Florida. Previously, he was a professor in the Department of Sociology at the University of Texas at Austin, where he was affiliated with the Population Research Center. He has held visiting appointments at the Center for Development and Regional Planning (CEDEPLAR) at the Federal University of Minas Gerais, in Belo Horizonte, Brazil. His areas of specialization include demography, Latin American studies, and the sociology of development. His research has focused mainly on the Amazon region of Brazil and the processes of land settlement and deforestation that have recently taken place in that part of the world. He has published three books and numerous scholarly articles. He received a PhD degree in sociology from the University of Texas at Austin.